THE
COMPLETE
IDIOT'S
GUIDE® TO

Evolution

DEMCO

Publisher
Marie Butler-Knight

Product Manager
Phil Kitchel

Managing Editor
Jennifer Chisholm

Acquisitions Editor
Randy Ladenheim-Gil

Development Editor
Tom Stevens

Production Editor
Katherin Bidwell

Copy Editor
Krista Hansing

Illustrator
Brian Moyer

Cover Designers
Mike Freeland
Kevin Spear

Book Designers
Scott Cook and Amy Adams of DesignLab

Indexer
Brad Herriman

Layout/Proofreading
Angela Calvert
Svetlana Dominguez
Gloria Schurick
Sherry Taggart

Contents at a Glance

Contents

Foreword

LUCA (Last Universal Common Ancestor) was born in a murky puddle. No matter. It happened roughly three billion years ago and LUCA was just a cell, or at least the beginnings of one. On the other hand, as the Last Universal Common Ancestor, LUCA gets credit for seeding all the life on our otherwise negligible little planet. Since then, the very game of chance and selective perpetuation that created LUCA has installed animal, plant, bug, and vegetable in virtually every nook and cranny of the world.

The game is called evolution and its rules were limned a century and a half ago by the brilliant English naturalists Charles Darwin and Alfred Russel Wallace. Their theories have since been tested, modified, retested, and refined to become the organizing principle of modern biology.

So what exactly does that theory say? It essentially says that the diverse life we see around us is a product of statistical inevitability. The argument is as simple as it is elegant. It holds that living things that adapt to the changes in their world are the ones most likely to survive and perpetuate their kind. Those that don't adapt die out—victims of what Darwin termed "natural selection." If the newly acquired traits of the adaptable are transferred to their offspring, the argument goes, those offspring are more likely to thrive, dominate their species, and eventually fuel its evolution. (Perpetuation is pretty much the name of the game: Longevity is only of evolutionary benefit if it aids procreation.) Ever since LUCA, this continual thrust and parry of adaptation and change—not some supernatural, moral force, or creative intelligence—has generated everything from aardvarks to zinnias. In short, evolutionary theory asserts that all living things are related.

What it explains—just as powerfully—is diversity. Ours is a world of oceans, deserts, prairies, jungles, bogs, reefs, and rivers, to name but a handful of habitable niches. A bewildering array of living things, shaped by natural selection, have adapted to exploit every conceivable resource in such places. More than a million species have thus far been identified, but that's hardly anything compared to the innumerable ranks of species we haven't yet seen or named. Such diversity is elegantly explained by the pull of evolutionary forces.

That's not all. Bolstered by advances in genetics and biochemistry, modern evolutionary theory elucidates phenomena as distinct as road rage and infectious disease. It accounts for why AIDS is so difficult to treat and why pharmacologists are still stumped by the common cold. In fact, the fundamentals of the theory remain so sound that scientists analyzing the human genome routinely use them to hone in on the molecular biology of newly discovered genes.

Many people have rejected the moral neutrality of evolutionary theory, troubled by where it leaves divinity, doctrine, and purpose in the scheme of human existence. If you're one of them, put this book away. Don't waste your time. This book isn't really for idiots, but then, neither is it for people who've already made up their minds to reject science when it challenges their beliefs. If, on the other hand, you're curious, get ready for a highly accessible discourse on the theory of evolution and its historical development.

—Unmesh Kher
Science Writer

Unmesh Kher was born in India, raised in Zambia, and attended college and graduate school in the United States. He holds Masters' degrees in molecular pathology and in journalism, both from New York University.

Introduction

> "Nothing in biology makes sense except in the light of evolution."
>
> —Theodosius Dobzhansky, American zoologist and geneticist

Although Darwin proposed his famous theory of natural selection more than a century and a half ago, evolution is still making headlines. Just as an illustration, here's a recent sampling taken from *USA Today* and *The New York Times*:

➤ "Darwin vs. Design: Evolution's New Battle"

➤ "Biology Text Illustrations More Fiction Than Fact"

➤ "Link Between Human Genes and Bacteria Is Hotly Debated by Rival Scientific Camps"

➤ "Kansas Restores Evolution to Science Standards"

➤ "Toolmaking May Have Influenced Human Evolution"

➤ "Fossil Causes Problems in Evolution Conflict"

What makes evolution such a controversial subject? Why have school boards divided over whether evolution should be taught in the classroom as an established science or simply as a theory of questionable validity? And why has the Supreme Court been drawn into the fray? No one, after all, is railing against the teaching of relativity or plate tectonics, which are also theories. What is it about the theory of evolution, then, that arouses such strong emotions—pro and con?

Well, as a start, consider this: The theory of evolution tries to answer two of the most profound questions of all time: How did life originate on Earth, and how has life developed since creation? Or, to put it in more personal terms, why are we—all six billion plus of us—here? Evolutionary theory does not pretend to answer those questions with absolute certainty. What the theory does do is try to explain how life developed on Earth and account for why some species, such as humans, have succeeded (so far, at least) in dominating the planet, while other species, such as dinosaurs, vanished long ago. As you'll see, there is a great deal more to evolutionary theory than Darwin.

Long before Darwin's birth, philosophers and naturalists were making valiant attempts to understand the underlying patterns of nature. But every time someone came up with a theory to account for the origin or evolution of life, new evidence would turn up—a marine fossil embedded in rock in the Welsh countryside, for instance, or the remains of a possible prehuman in East Africa—to send people running back to the drawing boards. It was as if you had nearly solved a puzzle but then were handed several more pieces and asked to fit them in as well.

This book is a story of how a great many very smart individuals kept picking up the pieces, trying to make them fit. You can think of this book as a journey—a chance to retrace the steps of some very remarkable thinkers, each of whom either solved a portion of the puzzle or contributed more "pieces" to it.

And this journey didn't come to an end with Darwin, either. He represents one very important stop along the way. Darwin built the foundation for what was to come, but, as he was the first to admit, his theory had a lot of holes in it. For example, he had no idea about the biology of inheritance, how traits are passed from one generation to the next—why you have your grandfather's temperament, your father's nose, and your mother's blue eyes. In this book, we'll see how other scientists picked up where Darwin left off. As you'll see, when you learn about evolution you're also learning about genetics, geology, paleontology (the study of fossils), anthropology, biology, religion, politics, and demographics, with a little bit of chemistry—and, yes, even sex—thrown in for good measure.

When we talk about the "theory of evolution" it's usually taken to mean Darwin's theory. And, for the most part, that's true because Darwin's theory has held up wonderfully for many years. But, in a larger sense, there are several different theories of evolution, and we'll be taking a look at some of the most important of them as well. As you go along, bear in mind that many of these theories are heatedly disputed among scientists. And it's not just the theories that are contested—so are the facts that the theories are meant to explain them.

As a philosophy professor once said, "The facts are never all in." That's certainly true when it comes to a subject as complex as evolution. Even those "facts" that are in are subject to conflicting interpretations. We aren't going to make judgment calls about which of them is "right" or "wrong"—usually there is a little of both. What we want to do is give you a sense of the lay of the land—what's behind all those headlines and heated disputes—so that you can grasp what evolution is all about and form your own opinions. Where you go from there is up to you.

To help smooth your way, I've included three sidebars, which appear in every chapter:

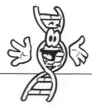

A Natural Selection

This features information of historical, geographical, or political relevance to the text.

Who Are They?

This sidebar offers brief biographical profiles of many of the important figures in the history of evolution.

What Does It Mean?

This sidebar provides succinct definitions of terms used in the chapter. (If you need additional help you can always turn to the Glossary for definitions of all the scientific terms in the book.)

Acknowledgments

To Jeanne Oliver whose kind hospitality in Nice made the writing of this book a more pleasant experience than it would have been otherwise.

Trademarks

All terms mentioned in this book that are known to be or are suspected of being trademarks or service marks have been appropriately capitalized. Alpha Books and Pearson Education, Inc., cannot attest to the accuracy of this information. Use of a term in this book should not be regarded as affecting the validity of any trademark or service mark.

Part 1
The Evolution of an Idea

How did life arise on Earth? Where did we—Homo sapiens—come from? Those are the kind of questions that have kept humans up at night for millennia—they still do. For that matter, how is it that we developed the intellectual capacity even to ask these questions (and suffer insomnia from them)? While we usually credit Charles Darwin with unraveling the mysteries of evolution, his achievements wouldn't have been possible without the contributions of many naturalists, geologists, zoologists, explorers, and philosophers who preceded him.

In Part 1, we see how the ancient Greek philosophers, such as Aristotle, struggled to come up with a theory that would account for how life originated on Earth. (After all, Aristotle had a theory for just about everything else.) Were all species present at creation? Did organisms change—that is to say, did they evolve or did they stay the same forever? For early Christians, on the other hand, there was no need to ponder such questions—the answers were there for all to see in the Book of Genesis. God created the universe and all species in six days and that was the end of the story. Evolution didn't enter the picture at all. But then geologists in the eighteenth and early nineteenth centuries began to uncover fossils of species that no longer existed anywhere on Earth. These discoveries raised some unsettling questions and no doubt induced more insomnia. Where had these ancient species come from—and even more puzzling—where had they all gone to?

Chapter 1

The Basics of Evolution

> ### In This Chapter
>
> ➤ What exactly is evolution?
>
> ➤ Why is evolution so controversial?
>
> ➤ The mechanism of evolution
>
> ➤ The contributions of Darwin and Wallace

What exactly is evolution? In a word, evolution is change. Evolution is a rational, comprehensible idea, based on physical evidence and experiments. It offers the best scientific explanation of our origins.

Evolution is one theory that has stood the test of time and is accepted by most scientists throughout the world. The word *test* is not a metaphor—every part of the theory of evolution is open to testing. And we have an almost infinite number of clues to draw from to see whether the theory stands up to the challenge, ranging from fossils to genes, to the similarity of flora and fauna on land masses separated by oceans. What scientists have discovered from examining all these clues is that all life is related to each other and to the *species* that came long before us and that have since vanished from the Earth.

However, because we refer to evolution as "a theory," many people have the impression that it may not be true or that a conflicting theory may be equally valid. Even people who do accept the idea that, to one degree or other, species undergo modification and are the descendants of earlier forms of life, they aren't able to tell you why.

Even for those who understand it, Darwin's theory of evolution isn't the obvious explanation for why species change, die, or turn into another species altogether. Our common sense tells us that we should be at the center of all life and not find ourselves stranded—literally—on the end of a limb of an amazingly complicated evolutionary tree.

Why Oppose Evolution?

According to a 1991 opinion poll, 100 million Americans believe that "God created man pretty much in his present form at one time during the last ten thousand years." A large majority of those polled saw no reason to oppose the teaching of creationism at school. Creationism is the belief that God created the world and everything in it all at once, and that nothing has significantly changed since. In the late 1990s, the Kansas State Board of Education ruled that school textbooks could no longer use the theory of evolution as the sole explanation for the origin of life—a decision that was rescinded two years later. Kansas was by no means alone in trying to oppose the teaching of evolution or diluting its impact. No fewer than 50 pieces of legislation have been passed in several states that have targeted evolution.

What makes the idea of evolution so threatening? After all, other once-controversial theories have been embraced and are routinely taught without any controversy. No one today gets up in arms because Nicholas Copernicus, the Polish astronomer, proposed that Earth revolves around the Sun, a humbling discovery to people of the sixteenth century, who had cherished the idea that Earth was the center of the universe. Galileo, who got into trouble with the Catholic Church for announcing that he had proven Copernicus right, no longer causes any furor, either.

A Natural Selection

Alabama legislators called for a note to be inserted into biology textbooks: "This book may discuss evolution, a controversial theory some scientists give as a scientific explanation for the origin of living things .... No one was present when life first appeared on earth. Therefore, any statement about life's origins should be considered as theory, not fact."

Something similar happened when Alfred Wegener, an astronomer and meteorologist, proposed his theory of continental drift—the idea that the continents were once all joined together and, over the eons, drifted slowly apart until they reached their present locations—was greeted with derision from geologists in the early twentieth century; today, however, his theory is the foundation of contemporary geology. Einstein's groundbreaking theories of gravity and light also received a skeptical reception when he proposed them, but both have been confirmed repeatedly in rigorous scientific experiments in the years since. Any physics textbook that doesn't include them would be next to useless. So why is evolution, which predates continental drift and Einstein's theories of gravity and light, still clouded by so much controversy?

Why Is Evolution So Misunderstood?

Because evolution is so poorly understood by so many, it's no wonder that its detractors can pull the wool over people's eyes. Why so much ignorance about a subject of such vital importance? (It has a personal impact—it explains how you got here.) For one thing, people suspect that evolution is too complicated to grasp without spending years plowing through impenetrable books. And because teachers tend to shy away from teaching the subject, if only to avoid upsetting parents, school boards, and religious organizations, many people never learned much, if anything, about it. In addition, most Americans aren't scientifically literate. According to a worldwide poll, Americans ranked twenty-first among industrialized nations on their knowledge of evolution! The blame for this miserable state of affairs doesn't lie only with nonscientists, either. At least some of the ignorance is due to misconceptions spread by people who should know better—the authors of dictionaries.

In the *Oxford Concise Science Dictionary*, for example, evolution is defined as "the gradual process by which the present diversity of plant and animal life arose from the earliest and most primitive organisms, which is believed to have been continuing for the past 3,000 million years." Only plant and animal life? What happened to protozoa and fungi such as mushrooms? Are they somehow exempt from evolution? Moreover, this definition leaves you with the impression that evolution was a process that occurred in the distant past and that it's all wrapped up and done—which isn't the case at all. *Webster's Dictionary* defines evolution as "the development of a species, organism, or organ from its original or primitive state to its present or specialized state." This isn't very accurate, either—at least, from a biologist's point of view—because it seems to suggest that evolution can occur within an individual (not true) and because it fails to account for how very complex creatures can evolve from other very complex creatures to form new *species*.

What Does It Mean?

A single **species** is defined as a distinct kind of organism, with a characteristic shape, size, behavior, and habitat that remains constant from year to year. In biological terms, a species is a natural population whose members mate and produce offspring only with one another, not with other populations.

Let's Get Our Terms Straight

We said at the beginning of the chapter that evolution is simply about change. But what kind of change? You can't begin to understand a concept, let alone dispute it, before you actually have a handle on what you're talking about. First of all, it's important to distinguish between the existence of evolution and various theories about the mechanism of evolution—how evolution actually works. What exactly do biologists mean when they say that they have observed evolution or that humans and chimps have evolved from a common ancestor?

An Expert's Definition of Evolution

This is how an eminent evolutionary biologist Douglas J. Futuyma defines evolution:

> In the broadest sense, evolution is merely change, and so is all-pervasive; galaxies, languages, and political systems all evolve. Biological evolution … is change in the properties of populations of organisms that transcend the lifetime of a single individual.

Notice the emphasis on populations. Evolution does not affect individuals, but only large numbers of people or groups—species. Futuyma then goes on to say:

> The changes in populations that are considered evolutionary are those that are inheritable via the genetic material from one generation to the next.

Biological evolution may be slight or substantial; it can involve changes in *genes* that determine blood types in a given population (slight) or the changes over millions of years that led to the development of humans and apes, as well as birds, mice, wasps, orchids, and daffodils (substantial).

Evolution Takes Time

Where evolution is concerned, as Futuyma points out, change cannot last only one generation (in which case, it would affect individuals); it must be a change that is inherited. Inheritance, as we know (and as Charles Darwin did not), is passed by genes. Even when evolution is gradual, it can still proceed fairly rapidly. In only a century, for instance, biologists noted a change in the frequency of genes in the American population. The average adult man and woman are significantly taller today than in the early 1900s. That's because, over the last several generations

What Does It Mean?

A **gene** is a unit of biological information, which, alone and together with other genes, determines the inheritance of various traits. In 2001, scientists announced that they discovered that the human has about 30,000 genes (much less than previously thought), while a mere roundworm has about 9,000.

of Americans, the genes for greater height have become more predominant in the population, probably as a result of better nutrition.

Where Did You Come From?

You can't have evolution, however, without something to evolve *from*. For instance, some 50 million dogs now reside in the United States alone. But for as many different breeds of dogs as there are, from St. Bernards to Pekinese, from Labradors to cocker spaniels, they all have one common ancestor: the wolf. Humans and chimps, too, evolved from a common ancestor before they went their separate ways. (Humans didn't evolve from apes or monkeys—that's a misconception.)

Defining evolution:

> ➤ Evolution is change.
>
> ➤ Evolution affects populations (or species), not individuals.
>
> ➤ Evolution may entail slight or significant changes to a species.
>
> ➤ If the changes are significant enough and continue over successive generations because they are favorable to survival in the environment, a new species may develop.
>
> ➤ Present-day species, however diverse, come from a common ancestor. All life is related.

A Natural Selection

According to Ernst Haeckel, a nineteenth-century German biologist and philosopher, "Ontogeny recapitulates phylogeny"—fancy words meaning that the development of the human embryo retraces the evolutionary development of the human race, each beginning with a simple organism (a cell) and evolving into complex forms.

Why Does Darwin Matter?

Whatever we think (or do not think) of evolution, we most commonly associate the theory with Darwin—and with good reason. But the idea of evolution did not begin with Darwin, who published his famous book—*The Origin of Species,* which he described as "one long argument"—in 1858. Evolution, as a concept, had been around for thousands of years before Darwin. Darwin's major contribution was to accurately describe how evolution actually worked. He understood several aspects of the evolutionary process that had eluded many of his brilliant predecessors. However, Darwin must share credit for his discovery with the English naturalist Alfred Russel Wallace. In one of the most famous coincidences ever recorded in the history of science, Wallace arrived at the idea of evolution practically at the same time as Darwin did. Darwin has been so closely identified with the theory of evolution, however, that his name will appear far more often than Wallace's in this book. But we'll try to give Wallace his due.

Even as late as the mid-nineteenth century, many people, including some eminent naturalists, still believed that species were immutable—that they had remained the same since creation and that each species was separately created—that is, had no common ancestor. (The question of when creation happened and *how* it happened was a subject of furious debate for hundreds of years.) Darwin, however, insisted that "species undergo modification, and that the existing forms of life are the descendants by true generation of pre-existing forms." Darwin also contended that the rules of the game remained the same: The natural and geological processes that are present now are the very same processes that were at work millions or even billions of years ago. If Darwin was right (and we think he was), evolution is a continual process, whose past can be deduced from studying the present.

What Does It Mean?

Simply put, **natural selection** is a weeding out process: Those organisms with traits that make them better able to adapt to a particular environment are more likely to pass those traits (in the form of genes) to the next generation than organisms with traits that do not favor adaptation.

Darwin's First of Two Great Ideas

Darwin had two great ideas. The first was the tendency of all organisms to increase in population at an unsustainable rate—for example, mosquitoes in the summer and humans at any time. The rate of increase is exponential ($2 \times 2 \times 2$), not arithmetical ($2 + 2 + 2$), which explains why we now have about six billion people on the planet (and far more mosquitoes) than populated the Earth only a few generations ago. But if the growth of any given population increases too much, Darwin said, something has to give. War, famine, pestilence, and the destruction of the environment are just some of the more catastrophic consequences that occur when the human population explodes. At some point, Darwin recognized, such explosive growth must be brought to a halt. (This is exactly what war, famine, pestilence, and environmental destruction do.) That recognition paved the way for Darwin's second great idea: *natural selection*.

Darwin's Second Great Idea

A struggle for existence is constantly raging, Darwin declared. The winners of the struggle are those that demonstrate better survival skills; they are better able to thrive in a particular environment, for instance, than their less resilient brothers and sisters. An organism with even the slightest advantage over its competitors—whether members of the same or a competing species—will have a better chance of passing its genetic legacy on to the next generation. This is what is meant by natural selection. Natural selection is the mechanism by which evolution works.

Let's look at Darwin's most important contributions to the theory of evolution:

➤ If their growth went unchecked, species would multiply beyond the capacity of the environment to sustain them.

➤ A struggle for existence goes on continuously between different species and within the same species.

➤ In the competition for resources, mates, and territory, some members of a species are eliminated. Species with traits that benefit them in this competition are more likely to survive. This is natural selection.

➤ Individuals who have traits that help ensure their survival are more likely to have offspring, which will inherit these same traits.

➤ The survival of the fittest takes place because of natural selection.

Genes and Inheritance: What Darwin Didn't Know

What makes Darwin's theory all the more remarkable is that he derived it without the benefit of understanding how traits are inherited. We now know that traits—eye color, height, temperament, and even a susceptibility to alcoholism or depression—are passed on by genes. Offspring are not exact copies of their parents. Each new generation of offspring will contain a natural range of genetic variation. That's another way of saying that we're not the same. Even identical twins have slight differences. Over successive generations, errors can creep into the mechanism responsible for transmitting genes. The result is *mutations*. Most of these mutations are harmless, but some can cause a great deal of damage: Juvenile diabetes, some forms of cancer, and multiple sclerosis are among the many diseases that are genetically linked. Sometimes, though, genetic mutations turn out to be beneficial for the survival of an organism.

Natural selective pressures—predators, abrupt changes in the climate or environment—can bring about a disproportionately high number of deaths of organisms that are less well adapted to their environment, while organisms fortunate enough to

What Does It Mean?

A **mutation** is a change in the structure of the genetic material (DNA) in a cell; mutations can occur because of damage to the cell—for instance, by exposure to radiation—or in cell division.

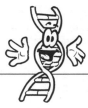

A Natural Selection

Darwin was not just an armchair theorist, but he also was an avid student of nature. He once cleared a plot of ground 3 feet long by 2 feet wide and counted 357 weeds. Within a short time, 295 were destroyed by birds and slugs. The lesson? Even pests can't run amok forever.

have acquired the beneficial mutation survive. As a result, those survival traits are passed on; in time, the population (of gnats, elephants, or humans) will be composed of more individuals with those beneficial traits. We say that those traits are "selected" by nature.

Evolution Is Observable, but You Might Not Like What You See

Evolution is always taking place. The problem for biologists (and for anyone with a healthy sense of skepticism) is that it's usually difficult to see it happen. A species can originate only once—and, when it does, humans seldom are present to observe it. Nonetheless, the effects of natural selection—the modification of a species from one generation to another—can easily be seen in domesticated animals. Long before Darwin came up with his theory of evolution, farmers and breeders of livestock, dogs, racehorses, and pigeons understood that it was possible to create new varieties of species with more desirable traits. In Darwin's day, for instance, breeders produced birds as distinct as the pouter, the runt, the turbit, and the rock pigeon. They are all birds, but they are different species of birds. They have identifiable characteristics that distinguish pouters from rock pigeons, for instance, and they are species because they can mate only with one another.

Ever since the early 1980s, we've had the misfortune of witnessing another example of evolution in progress: the HIV virus that causes AIDS. A virus undergoes evolution, following the same rules of natural selection that any other species does. Natural se-lection accounts in large part for why the HIV virus is so virulent and so difficult to conquer. For example, there are two principal types of HIV (1 and 2) and several sub-types, each one exclusively adapted to a particular population with its individual en-vironment, culture, and sexual habits. But scientists believe that all the different strains of virus owe their origin to a common ancestor that once inhabited Central Africa. Because the genetic structure of each type differs somewhat from other types, a vaccine against HIV-1 (no effective one exists at present) will be useless against HIV-2.

AIDS can be seen as a berserk copying machine that churns out copies of itself faster than its adversaries can come up with strategies to defeat it. The copies contain ever more mistakes, mixing up the genetic composition of new generations of virus for which the world is unprepared. Evidence suggests that HIV penetrated the human population several times before it exploded in the early 1980s.

Why did AIDS explode? Because the natural conditions were ripe for it—and humans had a lot to do with making it much easier for HIV to spread. War and economic tur-moil in Central Africa in the last several decades forced people to abandon their an-cestral villages and seek work and safety in the cities. When they migrated, the virus hitched a ride with them. In the cities, the familial and social structures that had worked so well in the countryside broke down; sexual promiscuity, intravenous drug use, and the means to travel quickly over great distances offered the virus an unparalleled opportunity to proliferate. Up until the 1980s, the virus had had only a handful of victims to choose from because it had been isolated in a remote rural environment; now the whole world is its oyster. If anyone has doubts about the existence of evolution, HIV should set them decisively to rest.

As we said earlier, neither Darwin nor Wallace was the first to introduce the concept of natural evolution to the world; what they did was refine it and explain it in a way that best fit with the facts. But to appreciate what these two great theorists achieved, we need to see how the concept of evolution itself evolved. In the next chapter, we will go back several thousand years to the era of the Greek philosophers who fearlessly confronted the greatest mystery of all—how did we get here?

The Least You Need to Know

➤ Evolution, in its most basic level, means change.

➤ The theory of evolution still offers the best explanation for the development of all organic life on Earth.

➤ Evolution occurs over time among populations, not among individual members of a particular group.

➤ All species, from the simplest one-celled organism to the most complex primate, undergo modification over long periods of time.

➤ Charles Darwin (and his fellow naturalist Alfred Russell Wallace) did not invent the idea of evolution, but they came up with the most accurate explanation of how it actually worked.

➤ Evolution is still going on, and it can be seen and verified by scientists.

Chapter 2

The Greek Philosophers Tackle Evolution

In This Chapter

➤ The Greek view of nature

➤ Anaximander: how life came out of the primordial ooze

➤ Aristotle and the ladder of life

➤ Empedocles and the grab bag of creation

➤ The Stoics and Epicureans: Defining the natural order

For the people of ancient times, the age of the Earth and the universe was as much of a mystery as how life originated on Earth. So the Greek philosophers, who were among the first thinkers to grapple with the idea of evolution, were working under two enormous handicaps. Even so, some of them came up with remarkable insights derived from direct observation and brilliant speculation, which revolutionized the way in which humans looked at their world.

As we'll see, the Greek thinkers approached these two handicaps in two basic ways. Some boldly dived into the question of how life originated and put forward their own creation myths. Others, notably Aristotle, decided to leave the whole question alone and simply address what could be said about the species that could be observed on Earth at the present time.

In addition, the Greek philosophers were at odds over another question that went to the heart of evolution and that was far from resolved for centuries afterward: Do species change and die out, with new ones emerging all the time, or are the species basically immortal, unchanging and fixed?

Anaximander: All Life Comes from the Sea

Born in 611 B.C.E., Anaximander conceived of an Earth in its infancy that was a mixture of mud and ooze. The wettest parts formed the sea and the drier parts formed the land, which was like potter's clay in consistency. As the Sun shone on Earth, Anaximander said, the heat produced delicate membranes that formed in the sea, swamps, and marshes: "And while the wet was being impregnated with life by reason of the warmth ... by night the living things forthwith received their nourishment from the mist that fell." When the embryos had fully developed, the membranes broke open, with the result that "there was produced every form of animal life." In Anaximander's scheme, the first generation of living things came about through a *metamorphosis,* in a process similar to the way in which a caterpillar is transformed into a butterfly.

What Does It Mean?

Metamorphosis refers to a process by which insects begin life in one form called larvae and undergo a dramatic change to another form. Such insects include butterflies, moths, beetles, bees, and flies. After the larvae grow to their full size, they turn into pupae, where their shape changes drastically within a protective shelter (it's called a chrysalis in butterflies, and it's called a chamber or cocoon in other insects). When it reaches maturity, the adult insect breaks out of the structure and flies off.

Living Things of Another Kind

Those creatures that had received the most warmth from the sun "set off to the higher regions" (became birds), while those creatures that retained "an earthy consistency came to be numbered in the class of creeping things and of other land animals." Similarly, "those whose composition partook the most of the wet element" became the "water animals."

As the Earth cooled and grew more solid, it became impossible to generate any of the larger animals from these membranes. Instead, "each kind of living creatures was now begotten by breeding with one another." That was how humans came into existence, said Anaximander. In a statement that sounds eerily prescient, the Greek philosopher said:

> Man to begin with was generated from living things of another kind, since, whereas others can quickly hunt for their own food, men alone require prolonged nursing. If he had been like that in the beginning, he would never have survived ….

An observer of nature, Anaximander was impressed by certain types of sharks or dogfish that struck him as having characteristics of both fish and land animals; he considered them as intermediaries between the two realms. Humans owed their origin to just such intermediaries—fish-like creatures—that nurtured human embryos until they had reached maturity. "Then at last the creatures burst open, and out of them came men and women who were already able to fend for themselves."

What Anaximander Got Right ... and Wrong

It does not demean Anaximander to say that his theories were, for the most part, preposterously wrongheaded—although his scenarios of creation might serve as an inspiration to writers of Hollywood horror scripts. He was struggling to explain the origin of all organic life on Earth with very little evidence to go on. That he believed that one species could give birth to another was, in its way, a brilliant insight. Where he missed the boat, however, was in his failure to realize that species could also undergo change. The alteration of species over time is a keystone of evolution, and that was something that Anaximander didn't figure out.

Aristotle: The Ladder of Being

Perhaps the greatest philosopher of all time, Aristotle took on the problem of organic change with the same boldness with which he confronted any problem that attracted his interest, whether biological, metaphysical, or philosophical. Rather than try to come up with a theory on how life began on Earth, as Anaximander had, he simply declared that there had never been a "beginning of all things," putting an end to any discussion on the subject before it could get started. The natural order, Aristotle contended, was eternal and unchanging, having neither a temporal origin in the past nor the capability of being destroyed in the future. Only individuals were born and died; species were eternal. Nature was, in a word, static. Aristotle was so dazzled by the tremendous variety to be found in nature that he surmised that every single kind of living creature that could exist was already present on Earth.

Who Are They?

Considered the greatest thinker in Western history, Aristotle (384–322 B.C.E.) systematically classified all extant branches of knowledge, including biology, psychology, physics, and literary theory. He also invented formal logic, developed zoology, and grappled with the major philosophical problems of his—and our—day. When writers in the Middle Ages referred to "the Philosopher," no one had any doubt whom they were referring to.

Where Do They Belong?

In his pioneering zoological studies, Aristotle classified the species of the animal kingdom into two main groups: *vertebrates* and *invertebrates*. He further subdivided the two groups according to the way offspring are produced—whether the organism was born alive, hatched from eggs, emerged from pupae, and so on. But he was an acute observer, and as he proceeded with his studies, he recognized that organisms didn't fall so neatly into the categories he had tried to cram them into. The borders separating his classes of creatures began to blur. To make things even more complicated, ambiguous species kept being discovered. These were "intermediate" forms of life, Aristotle decided. "The transition by which Nature passes from lifeless things to animal life is so imperceptible," he wrote, "that one can determine no exact line of demarcation, nor say for certain on which side any intermediate form should lie."

What Does It Mean?

Vertebrates are defined as animals with a segmented spinal cord in their adult space. Vertebrates include mammals (including humans), birds, reptiles, amphibians, fish, rays, and lampreys. **Invertebrates,** on the other hand, are defined as any organism without a backbone, including insects. Invertebrates constitute the majority of the animal kingdom.

Climbing the Ladder of Being

In his attempt to determine what to do with all these vexing intermediate beings, Aristotle conceived of one of his most important ideas—the ladder of being. At the bottom rung of this ladder were "the lifeless things." On the next rungs were to be found various classifications of vegetables, based on the degree of "vitality" that they possessed—"in short, the whole vegetable kingdom, while lacking 'life' if compared with animals, possesses a great deal of 'life' by comparison with other bodies …." In this manner, Aristotle concluded, "Nature passes from lifeless things to animals in an unbroken sequence, through a range of intermediate beings which are alive and yet do not represent true 'animals' …."

Empedocles: Limbs and Organs Up for Grab

Another important Greek philosopher named Empedocles (490?–430 B.C.E.) came up with a theory of organic change that foreshadowed Darwin's theory of natural selection. In fact, Darwin did acknowledge hints of Empedocles's ideas in his own work. But, as you'll see, there were several significant differences between the two.

Unlike Aristotle, who dismissed the question of the origin of life as more or less irrelevant, Empedocles postulated what might be called the grab bag scenario of creation. When life first appeared on Earth, he said, it spontaneously generated every conceivable combination of limbs and organs in a process that was entirely random. But not every part proved useful, and, over time, those parts fell by the wayside, consigned to the dust heap of history. Only the best-adapted of living things survived once the dust had settled and all the ill-adapted forms were eliminated. In the sense that this theory considered adaptation as an explanation for how organic change took place, Empedocles was on to something important.

When Empedocles turned to the question of how life began in the first place, he emphasized the role of accident and chance. "Why should not Nature work, not for the sake of something or because it is better so, but just in the way the sky rains—not in order to make the corn grow, but of necessity?" he wrote. Water evaporates from the surface of the Earth and then returns in the form of rain; because of the rain, corn grows in the fields. But, Empedocles said, this natural process of evaporation and rain wasn't designed to produce corn. Corn happened as an unintended result. "Why, then, should it not be the same for the origin in nature …?"

A Natural Selection

Empedocles was not only a philosopher, but also a notable statesman. After participating in the overthrow of the ruling oligarchy of his native Agrigentum (now Agrigento, in Sicily), he was offered the crown by the city's grateful citizens. Instead of accepting it, he established a flourishing democracy.

Our organs have turned out to be extremely useful for getting by in the world, he said, but that wasn't their original purpose. It just turned out that they were better adapted. In other words, success in the natural order was pretty much based on coincidence. If it worked, it stayed; if not, it was discarded: "[S]uch things survived, being organized spontaneously in a fitting way; whereas those which grew otherwise, perished and continued to perish."

So what was the major difference between Empedocles's theory and Darwin's theory of natural selection? For Empedocles (and for Anaximander), once you could explain how existing species got to where they are already, your job was pretty much done. Natural selection, in this view, was a one-time thing: Once the ill-adapted organs were out of the picture, there was no need for further weeding and culling. Darwin, however, regarded natural selection as a continuous process operating on varying populations of species over millions of years. Darwin wasn't interested only in explaining how present-day species had come into existence (which Empedocles said happened purely by chance). Instead, he was interested in lines of descent—how present-day species evolved from common ancestors and how today's species, in turn, become ancestors of species yet to develop.

The Stoics: Change As Fate

The Stoics were the most influential school of Greek philosophy that thrived in the Roman Empire before the Christian era. The Stoics believed that truth—especially when it came to leading an ethical way of life—was to be found in nature, not among humans. They were convinced that the natural order was based upon rational principles. Everything that happened in the world could be explained as a result of a deterministic system of action and reaction; there was a limit to what humans could do to change this system (which accounts for the fatalistic attitude that we associate with the word *stoical*).

They differed from Aristotle in that they did not believe that nature was fixed. Instead, the Stoics believed that there was a "cosmic rhythm," in which everything was caught up in a common, all-embracing cycle of death and rebirth. That is to say, organic life did change, but this change happened not because of chance or coincidence, but rather because of fate.

The Epicureans and the Blessed Phase of History

The Epicureans represent another school of philosophy that flourished in ancient Rome in the pre-Christian era. This philosophy, based on the teachings of the Greek philosopher Epicurus, put forth the idea that pleasure is the supreme good and major goal of life. (But don't sign up to join just yet.) By *pleasure,* the Epicureans were referring to the joy of intellectual pursuits rather than sensual pleasure, which, the Epicureans noted, tends to disturb peace of mind. (They were pretty smart folks.)

Like the Stoics, they also took an interest in the natural order. And, like the Stoics, they conceived of a system based on cycles, beginning with a Golden Age—a more "blessed phase of history," when men were stronger and life was easier. It was during this Golden Age that most creatures on Earth were produced, including humans. That was because the soil was more fruitful, the climate was better, and the Earth had secreted a kind of nutrient "milk." This milk allowed the first human infants to survive to maturity without the benefit of female nurses.

Lucretius: All Change Comes to an End

The best-known Epicurean of Roman times was the poet Lucretius (94?–54? B.C.E.), who believed that living things came into existence at some definite moment in the past and that they did so by natural processes. Once on Earth, though, all change came to an end. Species were immutable and further evolution was impossible. Here is how Lucretius described the process in his famous poem, "On the Nature of Things":

> [E]ach thing has its own process of growth;
> All must preserve their mutual differences,
> Governed by Nature's irreversible law.

The Race Belongs to the Swiftest

But Lucretius did believe in a form of natural selection. Looking back on the Golden Age, he wrote:

> For we see numerous conditions first
> Must meet together, before living things
> Can beget and perpetuate their kind ...
> And many breeds of animals in those days
> Must have died out, being powerless by their offspring
> To perpetuate their kind.

Those creatures that survive to the debased age in which Lucretius lived (and presumably the age in which we, too, reside), must have had a definite advantage:

> For all those creatures
> Which now you see breathing the breath of life,
> 'Tis either cunning, or courage, or again
> Swiftness of movement, that from its origin
> Must have produced and preserved each race.

The Legacy of the Ancient Greeks

So, in our brief survey of ancient Greek philosophical attitudes about organic change, we've seen quite a variety of divergent viewpoints. To account for organic change (or its lack thereof), these brilliant minds had to struggle with the mystery of the origin of life. How do you get from there (the beginning) to here?

As we've seen, Aristotle dispensed with the question altogether. What you see is what you get: The organic life that's present on Earth was always there and always will be. Anaximander and Empedocles begged to differ. No, they said, species had changed since they first appeared on the planet, but all those changes have already taken place. The Stoics, on the other hand, declared that organic change would occur in the future but that the change was based on eternally fated cycles. The Epicureans allowed for change based on a kind of natural selection; some species—"prodigious monsters," Lucretius called them—never survived the Golden Age, but the creatures of sufficient cunning and swiftness of movement that managed to do so are here to stay. No further evolution awaits them.

Many of the ideas that emerged from the Greek philosophers—Aristotle's ladder of being is a prime example—were to exert a lasting influence over subsequent efforts by naturalists in Europe to arrive at an explanation of how nature had developed since creation. But before we can get to that part of the evolution story, we first need to explore an even more powerful influence over Western ideas about nature than the Greeks. You've probably guessed it: It's the Bible.

The Least You Need to Know

➤ Several ancient Greek philosophers made valiant efforts to understand how organic creatures first appeared on Earth and how they developed into their present forms.

➤ Two major views of the natural order emerged from the Greek philosophers—organic life had changed or disappeared since its creation, or all life that has ever existed is present on the Earth today.

➤ Some Greek philosophers developed a primitive concept of natural selection.

➤ No Greek philosophy stated that existing species could undergo change; if change had occurred in nature, it was all in the past.

Chapter 3

The Age of Earth and the Fixity of the Species

> ### In This Chapter
>
> ➤ The early Christian view of nature
>
> ➤ Reconciling Genesis with the Greek philosophers
>
> ➤ Descartes: taking God out of the picture
>
> ➤ John Ray: deciphering God's master plan
>
> ➤ The sacred history of the Earth

The Bible had a profound influence over the progress of scientific inquiry until well into the nineteenth century. The idea that the Bible was literally true—that God created the heavens and the Earth in seven days, as recounted in Genesis—was passionately championed (as it still is among some fundamentalist religious groups today).

Needless to say, that position didn't leave much room for scientific advancement. Many people, of course, while accepting the spiritual and moral teachings of the Bible, believed that the creation story was to be taken metaphorically, not as the literal truth. Other men of science (few women were permitted into their ranks until about a century ago) wavered between these two positions. They thought that there must be something to the story of the Flood and that this catastrophe must account for some of the geological and natural changes that occurred over the course of many centuries. But how many centuries, exactly?

Well, there was the rub, as they say. Apart from the impossibility of traveling back in time, how could anyone determine how old the Earth really was? And if you couldn't establish the age of the Earth, it would be hard to determine how species developed over time (if they did at all).

Over the course of several centuries, the Catholic Church relaxed its restrictive attitude about scientific investigation, as long as it served a spiritual or moral purpose. With the Protestant Reformation in the sixteenth century, however, the pendulum swung back toward fundamentalism. Genesis became the only authoritative account of creation. The timeline of the Earth again shrank to a matter of 5,000 years, and the planet's life expectancy shrank to a few centuries.

Scientists of the sixteenth and seventeenth centuries had to figure out some way of getting out from under the restrictions of religious dogma. Was there some clever strategy, they wondered, that would allow them to undertake their research without the church hovering over their shoulders? As it turns out, there was.

The Problem of Genesis

The Greeks, as we've seen, believed in the study of nature. From direct or empirical observations of nature, they maintained that it was possible to deduce certain rules or laws. Although they were mistaken on many counts, their approach to the subject is similar to what we refer to as the scientific method.

When Aristotle decided that "intermediary" forms existed because they shared characteristics of two different species (fish and animals, for instance), he was making an observation from nature, which served as the basis for a theory. The theory basically said that you would find intermediary forms on the ladder of being wherever you look. But for Christians, who held that the creation story of the Old Testament was literal, it wasn't necessary to examine nature (flora, fauna, rock formations, and so on) to discover the laws of nature. For that, you didn't have to look farther than the pages of Holy Scripture. Until 1000 C.E., philosophy had to take a back seat to theology.

However, many Christian thinkers were not so ready to abandon the contributions of Greek philosophy. On the contrary, efforts to reconcile *Genesis* and Greek philosophy had begun even before the birth of Christ. But it couldn't be done out in the open—anything that smacked of paganism was suspect because it called the truth of the Bible into question. As we'll see, though, some very clever Christian scholars found a way around the restrictions.

What Does It Mean?

Genesis comes from the Greek words *Genesis kosmou,* meaning "origin of the cosmos." It is the title of the first book of the Old Testament and the Five Books of Moses, known as the Pentateuch. The Hebrew text refers to Genesis by the first word in the book—*bereshith,* meaning "in the beginning."

Getting to Nature Through the Back Door

Throughout the Middle Ages, the Old Testament was considered an authoritative account of the cosmic history. The most popular version of this history comes from the third-century historian Julius Africanus, who, in a book titled *Chronographia,* sought to describe the events from the first day of creation until 221 C.E.

In his interpretation, the whole of history could be measured in terms of a cosmic "week," with each day consisting of a thousand years. By viewing history through the lens of the Bible, Christian scholars of the Middle Ages began to search through both the Old and the New Testaments for allusions to animals, plants, planets, and even geometrical forms in which they claimed to find hidden significance. Every serious biblical student of the time was expected to demonstrate adroitness in finding these elusive references and explaining them.

Does this all sound hopelessly esoteric? Actually, this wasn't as much of a meaningless exercise as it may sound. By employing a Church-sanctioned method of studying the Bible, scholars were able to conduct an investigation of history and nature. So, for example, if you wanted to learn more about things geological, you had only to find a biblical reference to rocks to justify it. In effect, using allusions from Scripture in this way allowed the intellectuals of the day to disguise their interest in the body of scientific and speculative work compiled by the pagans. Think of it as getting into the study of Greek philosophy through the back door of the Bible!

The great advantage to using biblical allusions was that you could say just about anything you wanted concerning nature, as long as it was cleverly camouflaged in theological terms. That is to say, exhibiting an interest in astronomy or zoology was no longer considered an obstacle to attaining salvation as a good Christian. One result of this approach was the popularization of medieval *bestiaries*—moral tales featuring animals that combined elements of natural history with theology.

What Does It Mean?

Bestiaries refer to a type of book that describes all the animals in creation (including some, such as the unicorn and the Phoenix, which didn't exist). These animals became the basis for allegorical tales that were used to promote Christian moral and religious instruction. Many artists, sculptors, and architects of the Middle Ages drew their inspiration from the symbolism in these bestiaries.

Take, for example, a bestiary tale of the industrious ant. When this creature needs food, we learn, it acts for the benefit of all its fellow ants. It has the responsibility of carrying one grain in its mouth. The author goes on to say, "Their comrades do not say 'give us of your grains' to the loaded ones, but they go … to the place where they have found their own, and carry back their own grain to the nest." In case you haven't drawn the proper moral from the tale, the author is happy to provide it: "Mere words, you see, are not an indication of being provident. Provident people, like ants, betake themselves to that place where they will get their future reward."

Reviving the Ladder of Being

Once Greek philosophy was considered respectable, Christian theology appropriated Aristotle's ladder of being and reinterpreted it for its own purposes. Remember that Aristotle had proposed a hierarchy of life, beginning with the simplest forms of lesser "vitality"—such as plants—rising rung by rung to ever more complex forms, with humans at the very top.

Christian theology took the concept one step—or rung—further: The ladder was no longer only a zoological classification system, but it was transformed into a chain of command. That meant that creatures occupying the higher reaches of the ladder were also superior to the creatures below them and so possessed more authority. God and his angels stood at the very top and held the most authority of all. God was the commander-in-chief. Humans retained a privileged position on Earth, which meant that they had a "divine right" to domesticate or slaughter animals as they chose.

The same rule applied all along the chain of command: The higher animals dominated the lower ones. Earth, being lifeless, was at the very bottom of the ladder; plants, rooted in earth, were higher up; worms and insects, which lived on the surface of the plants and the earth, were higher still; and quadrupeds (four-footed creatures) held an even more superior position. But humans alone, these theologians contended, occupied such a privileged place in this scheme because erect posture symbolized their spiritual striving to rise upward toward heaven.

In spite of the Christian spin given to the ladder of being, it still adhered to the old Aristotelian idea that the species were fixed: They had their place in the chain of command—and that was that. But not every thinker in the Middle Ages was so ready to assume that the species were fixed. Albert the Great, one of the most influential thirteenth-century scientists, believed that some kinds of living creatures (notably plants) could change into one another. Over the next several centuries, the possibility of organic change was a legitimate topic for discussion. It wasn't until much later, after the Renaissance, that the idea of the fixity of the species took hold in its most dogmatic form.

Turning Back the Clock: The Protestant Reformation

By the early sixteenth century, Catholic allegorical interpretation had succeeded in diminishing the literal belief in the biblical time scale. That all changed with the Protestant Reformation, led by Martin Luther. This new theology contended that the Catholics had strayed from the truth and that the old order had to be reestablished to bring about a "purer" Christian tradition. The new fundamentalism reasserted the belief that Genesis was to be taken literally, as the supreme authority on cosmology and geology. In this system, there was no room for allegory or allusion. For Protestant scientists, especially in England, God held a central place in the universe. By studying nature, the devout scientist was actually deciphering the purpose of creation and was trying to come closer to God.

Who Are They?

Martin Luther (1483–1546) was the founder of the Protestant Reformation. A German theologian who believed that God interacts with humans through the law and the Gospel, he is one of the most influential figures in history.

The Deluge Revisited

Protestant theology viewed the biblical Flood as the most powerful instrument of geological change. That was why the Garden of Eden no longer existed—the Deluge had washed it away. With the Fall of Man, corruption had taken root everywhere. "All creatures, yea, even the Sun and the Moon, have as it were put on sackcloth," declared Martin Luther.

Luther restored the old timeline, which held that 4000 B.C.E. was the date of creation. According to Luther, history was running out of steam: Humans were already in the sixth and final age, which had only another 400 years to go. The gig would be up by 2000. But Luther was skeptical that anyone would be living at the time. The corruption in the world was so pervasive that he couldn't imagine mankind lasting another four centuries. "The world will perish shortly," he declared. "The last day is at the door, and I believe the world will not endure a hundred years."

Earth Celebrates a Birthday

The 4000 B.C.E date favored by Luther as marking the beginning of creation was only one of several alternatives: 4032, 4004, 3949, and 3946 were all proposed at different points. Then the famous astronomer Johann Kepler tried to solve the question by examining the cycles of solar eclipses and comparing them with the New Testament dating of the Crucifixion. He found an error of four years, meaning that Christ's birth actually occurred in 4 B.C.E. It was this calculation that the Irish prelate Archbishop James Ussher used to set the date of creation at 4004 B.C.E.

Who Are They?

Archbishop James Ussher (1581–1656), the biblical scholar, was born in Dublin and was ordained an Anglican clergyman in 1601. In a two-volume work, entitled *Annals of the World,* he calculated that creation must have occurred in 4004 B.C.E, a date that was accepted for many years thereafter.

Who Are They?

René Descartes (1596–1650) is considered the first modern philosopher. He was always searching for certainty: How can we discover any truth? He conceived of a system of geometry, based on axioms and mathematical propositions, which were consistent and valid for all time and all places.

That was thought to settle the matter. The Bible said that Earth was more than 5,000 years old and was unlikely to live to see its 6,000th birthday. Not only that, but the species hadn't changed in those 5,0000-odd years, and no change could be expected in the few hundred remaining years left. As late as 1730, many European scientists still believed in a biblical chronology and species untouched by change.

Taking Creation Out of the Picture

If science continued to be subservient to religion, not much progress was likely to be made in studying the natural order. Enter René Descartes, the French philosopher who gave us calculus and the immortal words: "I think, therefore I am." When it came to unraveling the secrets of nature, Descartes decided not to get bogged down in the sensitive question of how the Earth and life came into being:

> There is no doubt that the world was first created in its full perfection, … there were in it a Sun, an Earth, a Moon, and the stars; plants themselves; and Adam and Eve were not born as babies, but made as full-grown human beings ….

As far as Descartes was concerned, Genesis was right on the money, and no one could ever accuse him of overturning accepted wisdom. But having established his theological credentials, Descartes then pulls off a neat little trick. "Nevertheless," he continues, "in order to understand the status of a plant or man, it is far better to consider how they may now gradually develop from seeds, rather than the way they were created by God at the beginning of the world …."

So we can see that Descartes was quite happy to set aside the question of creation altogether to concentrate on the present. It was *how* things developed, not *why*, that interested him. God was still in the picture, but there was no need to call on him to explain anything about nature as it is now. It is possible, Descartes declared, to discover the principles that govern nature by the light of reason alone. How could we be certain that our reason would give us the truth? Well, Descartes said, we know that because God endowed us with reason. So, paradoxically, God wasn't necessary to investigate nature, but He guaranteed that what was discovered about nature—so long as reason was applied—was going to be true.

What Descartes had done was to give scientists a sly but convenient way of wresting loose the hold that religious dogma had on scientific study. No longer was it necessary for scientists to engage in strenuous debates over whether it was actually possible to fit two of every creature on board Noah's Ark. Scientists could actually get down to the work of investigating nature—at least as long as they didn't start digging too far into the past.

John Ray: Searching for a Divine Master Plan

Like so many troubled marriages, science and religion didn't part ways so easily, or all at once. On the contrary, some scientists believed that by studying nature, they would be actually serving a theological end. The natural world offered an opportunity to discover evidence of God's master plan. This was the special objective of seventeenth-century English naturalist John Ray, who stood agape at the sheer abundance of species he encountered.

In his work *Wisdom of God* (described as "a scientific sermon"), Ray made an effort to classify 130 species of animals, snakes, birds, and fish. In addition, he estimated that there were around 10,000 species of insects and 18,000 of plants. "What can we infer from all this?" he wrote. "If the number of Creatures be exceedingly great, how great, nay, immense, must needs be the Power and Wisdom of him who form'd them all!"

Where Do Plants Belong?

Even as he was finding expression of God's presence in such a vast inventory of creatures, Ray was making a significant contribution to science. In 1682, he introduced his *taxonomic* system, in which he proposed that species were the basic unit of botany. Ray's is regarded as the first classification system of natural life.

In the 1690s, Ray reached two important conclusions from his classification of plants that had implications for other forms of life. First, he said, the essential characteristics of each species of plant were determined by the seed from which it grew. Second, a plant of one species never grew from the seed of another. Ray now had a criterion for distinguishing "real kinds" in nature. Two individual plants belonged to the same species if the one could be grown from a seed of the other or if they shared a common ancestor—they were related. This raised a problem, though, because how could

What Does It Mean?

Taxonomy refers to a system of classifying organisms. John Ray is considered the first taxonomist, but the greatest taxonomist of all is probably the eighteenth-century Swedish naturalist Carolus Linnaeus.

you definitively prove that two plants of different species hadn't been related at some point in the distant past?

John Ray's Dilemmas

Although most naturalists accepted Ray's theories about botany for the purpose of classifying different plants, he came in for some sharp criticism, too. It was all well and good, his detractors argued, for Ray to exult in "the fecundity of the Creator's Power," but how could he be so sure where types of plants as different as herbs and shrubs fit into his taxonomy? The ladder of being, they pointed out, was made up of a succession of indiscernible gradations between every step. So, once again, the old bugaboo of Aristotle's "intermediate forms" rears its ugly head! Did that mean that any system of classification would necessarily be arbitrary if you couldn't figure out what to do with these intermediate forms?

If trying to sort out present-day plant life wasn't hard enough, Ray confronted another dilemma when he studied fossilized traces of extinct ferns. Where did they belong? In a letter to a friend he wrote:

> ... there follows such a train of consequence, as seem to shock the Scripture history of ... the World; at least they overthrow the opinion generally received & not without good reason, among Divines and Philosophers, that since ye first Creation there have been no species of Animals or Vegetables lost, no new ones produced.

Suddenly the whole notion of the fixity of the species is thrown into doubt.

The Mystery of the Stones

That wasn't the only unwelcome discovery that Ray made in his quest to detect the divine master plan in nature. He had another surprise in store. When Ray went roaming around the valleys of Wales, he couldn't help observing the large number of boulders that lay strewn all about him. "I gather that all the other vast Stones that lie in our mountainous Valleys have by ... accidents fallen down." But how did they get there? Ray wondered.

Well, it was possible that Noah's Flood was responsible, but it was highly unlikely that even such a catastrophic deluge had reached all the way to the Welsh countryside. He noted, too, that these boulders must have been there for a very long time:

> For considering there are some thousands of them in these two valleys ... whereof there are but two or three that have fallen in the Memory of any Man now living; in the ordinary Course of Nature we shall be compelled to allow the rest many thousands of years more than the Age of the World.

Now there was an unsettling thought! The age of the Welsh boulders was older than the accepted age of the Earth upon which they lay. How was that possible?

The Sacred Theory of the Earth

Ray wasn't the only naturalist who hoped to keep religion and science bound together. In the 1680s, another Englishman named Thomas Burnet published a four-volume work titled *Sacred Theory of the Earth* in which he declared that the words of God (enshrined in the Bible) had to be consistent with the acts of God (exhibited in nature.) Burnet was determined to retell world history by means of traditional Bible stories, only with a novel twist: These stories were reinterpreted in naturalistic terms.

> Thus the Flood came to its height; ... when the Earth was broken and swallowed up in the Abyss, whose raging Waters rise higher than the Mountains ... and with thick Darkness, so as Nature seem'd to be in a second Chaos; and upon this Chaos rid the distress'd Ark, that bore the small Remains of Mankind ... a Ship whose Cargo was no less than a whole World that carry'd the Fortune and Hopes of all Posterity ....

Critics called the *Sacred Theory* fiction, dismissing it as a "historical romance." The bad reviews didn't hurt sales, though. It attracted every bit as much interest as Newton's monumental *Principia,* which laid the foundations of modern physics and continued to be a best seller for much of the eighteenth century. But Burnet's efforts to harmonize scripture with nature represented a last-ditch effort to keep science from breaking with theology. His attempts to explain the evidence of the epochal geological change by means of the Flood only succeeded in showing that nature and scripture weren't so easily harmonized.

Even some of the most devout Protestant scientists were beginning to question whether the constraints of dogma might be getting a little too tight. Maybe it was time to go out in nature and find out what secrets she might be coaxed into surrendering.

The Least You Need to Know

➤ In the Middle Ages, scientific investigation was permissible as long as it had a spiritual or moral purpose.

➤ The use of biblical allusions, allegories, and bestiaries allowed scientists to incorporate many of the ideas about creation and species from the Greek philosophers into Christian doctrine.

➤ Science suffered a blow when the Protestant Reformation ushered in an era in which Genesis was the ultimate authority on creation and the origin of the species.

➤ In Protestant theology, the age of the Earth was set at about 5,000 years, and Noah's Flood was considered the most powerful influence of geological change.

➤ The great seventeenth-century French philosopher René Descartes proposed that scientific investigation into the development of species could continue without repudiating the biblical story of creation.

➤ The eighteenth-century English naturalist John Ray sought to detect God's master plan in nature, but he began to realize that the master plan was far more confounding than he had ever imagined.

Earth Gets a History

Throughout the seventeenth century and well into the eighteenth century, there were only two games in town: Aristotle's ladder of being (in a revised form) and the Book of Genesis. The Earth was barely 5,000 years old, all evidence of dramatic geological change could be explained by the Flood, and the species that existed had existed since creation. But just underneath the surface—literally!—things were waiting to be discovered that would soon turn the received wisdom of two millennia on its head.

Credit for producing this upheaval belongs to the geologists. Not that they had any intention of exposing facts that would prove so devastating to the biblical timeline or the idea that the species were static, but that is what they did in spite of themselves. How to explain fossils of marine life in rocks on dry land? What was marine life doing on land in the first place, and where had the water gone? And what connection did these fossils have with existing species?

Even more disconcertingly, sometimes geologists found organic fossils that, while bearing startling similarities to existing species, also had very different characteristics from their modern descendants. It was harder still to explain away organic fossils that seemed so unusual that no present-day species resembled them at all! Nor could these bewildered geologists ignore the evidence of rocks, valleys, and oceans that argued in favor of a succession of great calamities rather than a single deluge recorded in the Old Testament.

When the evidence proved overwhelming in favor of geological change had been accepted, it raised new questions, no less unsettling. How did the change come about? How much time did it take? What forces were responsible, and were they still at work today? Each question required more research and exploration to answer, and each successive discovery, of course, only prompted more questions. But we seem to be getting ahead of ourselves.

The Chain of Being

By the seventeenth century, the ladder of being had acquired a new, and jazzier, name—the great chain of being. In what is considered the best-known statement describing the chain of being, the English poet Alexander Pope wrote:

> Order is Heav'n's first law; and this confess'd,
> Some are, and must be, greater than the rest,
> More rich, more wise ....

Who Are They?

Antoni van Leeuwenhoek (1632–1723), had worked as a haberdasher and a chamberlain for sheriffs of his native town of Delft, Holland, before becoming a scientist. As a hobby, he invented the microscope, which allowed him to make pioneering discoveries about protozoa, red blood cells, capillary systems, and the life cycles of insects.

So now you can see that the chain of being (Aristotle's ladder in a new package) was being used not only as a means of elucidating the natural order but also in justifying a certain social order—one that obviously benefited the rich and powerful. The phrase "laws of nature" was taken very seriously at this time because these same laws were thought to apply to civil society as well. Violating the laws of civil society—a criminal act or an act of revolution, for instance—also threatened the order of nature. You can see how kings, oligarchs, and religious authorities would find the idea of the chain of being such an appealing one.

If anything, the chain of being had expanded considerably since Aristotle's day, especially at the very bottom rung of the heap. Discoveries by the Dutch scientist Antoni van Leeuwenhoek, using a microscope that he had invented, had revealed a world of tiny creatures whose existence no one had previously suspected. "Every part of Matter is peopled; every green Leaf swarms with Inhabitants," wrote the

English essayist Thomas Addison, in astonishment. "There is scarce a single Humour in the Body of a Man, or of any other Animal, in which our Glasses do not discover Myriads of living Creatures ...."

The great British philosopher John Locke wrote:

> In all the visible corporeal world we see no chasms or gaps. All quite down from us the descent is by easy steps .... And when we consider the infinite power and wisdom of the Maker, we have reason to think, that it is suitable to the magnificent harmony of the universe, and the great design and infinite goodness of the architect, that the species of creatures should also, by gentle degrees, ascend upwards from us towards his infinite perfection, as we see they gradually descend from us downwards.

Such a view dominated thinking throughout the eighteenth century and strengthened the view that species were fixed and unchanging.

As long as the static picture of the universe held sway, the moment of truth was going to be put off. Sooner or later, though, scientists would have to confront the painful reality that Earth was much older than anyone had thought. Until this point, scarcely anyone had conceived of the possibility of nature changing through time. However, change was in the air—or, rather, in evidence that geologists were unearthing from the ground.

Mr. In Between

In the years following 1650, the ladder of being underwent another change to meet the needs of a new age. God was no longer the commander-in-chief at the pinnacle of the ladder, but He had changed his role, becoming the supreme designer of an elegantly wrought world-machine.

The problem of the "intermediate beings" underwent a redefinition, too. Previously, these beings were ambiguous creatures that had proven so difficult to classify—were they, literally, fish or fowl? Wrote John Locke:

> There are fishes that have wings and are not strangers to the airy regions; and there are some birds that are inhabitants of the water, whose blood is as cold as fishes .... There are animals so near of kin both to birds and beasts that they are in the middle of both ....

DARWIN

Who Are They?

John Locke (1632–1704) believed in the importance of experience of the senses for gaining knowledge rather than relying on intuition or speculation. His philosophical approach is known as empiricism. He viewed the mind of a person at birth as a blank slate—a *tabula rasa*—upon which experience imprinted knowledge.

But by the end of the seventeenth century and beginning of the eighteenth century, intermediate creatures meant something different.

The question about intermediate creatures was given a new and faintly ominous twist. What, naturalists wondered, was the missing link between rational and nonrational creatures? Surely, the thinking went, some form of primates must exist between the great apes and the most stupid humans. Jean-Jacques Rousseau, the eminent French philosopher, put the question like this:

> The Ape or the Monkey that bears the greatest Similitude to Man, is the next Order of Animals below him. Nor is the Disagreement between the basest of Individuals of our species and the Ape or Monkey so great, but that, were the latter endow'd with the Faculty of Speech, they might perhaps as justly claim the Rank and Dignity of the human Race, as the savage Hottentot ….

There's a time bomb ticking away in these words. That's because Rousseau's words carry the implication that certain races or groups of humans, such as Hottentots of southern Africa, which are more closely related to apes—and thus, have lower intelligence—than other humans. Later on, in chapters on social Darwinism and eugenics, we'll see just how poisonous these ideas actually are. Trying to sort out the question in theory as to where species belong in the great chain of being had the potential to do great damage in reality.

The Discovery of the Earth

What were the geologists actually digging up? For one thing, they were finding fossils of marine and animal life. Fossils were nothing new. Even in antiquity people knew about them, but they didn't attach much significance to them. They were curiosity items, nothing more. For another thing, geologists were beginning to explore, to a greater extent than before, the nature of the surface of the Earth itself. How had its mountains arisen and fallen? How had its valleys been carved out and its oceans been formed? Could all the evidence of dramatic geological transformation in the past be attributed to a single flood of biblical proportions?

Ironically, the first modern geologists didn't set out to explore the composition of the Earth's crust. They had no interest in classifying different rocks, nor were they curious about finding out whether any consistent order could be found in the strata of different countries and regions. To the contrary, many of them were more concerned with making a living by applying their skills to practical enterprises such as mining and metallurgy. Who could blame them? They had no reason to think of studying the Earth's history when they were virtually unaware that the Earth had any history at all.

Ah, but there were those damned discoveries that kept piling up, and they demanded answers. For instance, fossil marine shells, found in inland rocks, were difficult to explain by Genesis alone.

The Great Fossil Hunt

Two leading geologists of the seventeenth century—Robert Hooke, from England, and Nils Steensen, from Denmark—were among the first to recognize the crucial significance of organic fossils. Fossils, they determined, were of the same age as the rocks in which they were found. Why was that so important? Because it meant that if you could date the former, you could date the latter and, in addition, discover a great deal more about the history of the Earth and the species that had inhabited it.

"A great part of the Surface of the Earth hath been since the Creation transformed and made of another Nature; namely, many Parts which have been Sea are now Land; and diverse other Parts are now Sea which were once a firm Land," wrote Hooke. That finding suggested that some creatures found in rock on land might once have swum in seas long since receded. What's more, the two observed that many of the petrified organisms that they were collecting resembled organisms that existed in the present.

Unwelcome Discoveries

Not all of these discoveries were welcome. Like Ray, Hooke and Steensen had to contend with an embarrassing dilemma. The cataclysms responsible for shaping the Earth's surface must have been vastly more drastic and occurred far more frequently than any that they were accustomed to seeing in their own experience. That was the only way to explain how, for instance, a seabed could have been dislodged from its earlier location and why a valley had come to supplant it. But how could these calamities have come at relatively frequent intervals when, as far as the geologists were aware, only one significant catastrophe had occurred since Creation, and that was, of course, the Flood. At least, that is what all the available histories—and the Bible—said.

Well, perhaps, in an earlier era, the Earth's crust was far more malleable than it was in the present so that it was more susceptible to being reshaped—like a pot still in the kiln. That didn't solve the problem, though, because it left the question of why, if there were all these catastrophes, there was no record of them even in ancient times? And that wasn't the only difficulty that the geologists faced. Not all the fossils that they were unearthing resembled creatures still in existence. Instead, the fossil evidence suggested that many species in earlier ages consisted of creatures that no longer existed. What did that say then about the fixity of the species?

Then fossils turned up on the south coast of England that were, in Hooke's words, "of a much greater and gigantick standard; suppose ten times as big as at present." What had happened to them? Hooke decided that they still existed, only that the living descendants of these gigantic creatures had "dwindled and degenerated into a dwarfish progeny." He was struggling not to abandon the biblical time scale at all costs. Although his findings were indicating that the Earth had to be many thousands of years older than scriptural interpretation allowed, he couldn't take the necessary step to reconcile them with his beliefs. In fact, after being converted to Catholicism and receiving an appointment as bishop, he abandoned his interest in geology altogether.

35

Johann Herder and the Power of Nature

Although he was not a geologist, eighteenth-century German philosopher Johann Gottfried Herder put together an influential account of natural development in a four-volume opus called *Ideas Towards a Philosophy of the History of Man*. In it, he offered a compromise between the old and the new. First he reminded readers how the creatures in Aristotle's ladder of being had appeared in succession, occupying one rung after another all the way to the top, where the most complex and superior forms of life (namely, us) resided:

> From stones to crystals, from crystals to metals, from these to plants, from plants to animals and from animals to man, we see the form of organization ascend; and with it the powers and propensities of the creature become more variant, until finally they all, so far as possible, unite in the form of man.

Where Herder broke with his predecessors was in his assertion that each species changed an essential part of the environment for those which came later. The great chain of being—the seventeenth-century version of Aristotle's ladder—had previously been unchanging. The species didn't change in their nature, nor did they change their position on the ladder (or in the chain). Herder introduced a dimension of time: The lower creatures weren't unconnected to the species above them; instead, they paved the way for the higher creatures that would follow them. What Herder had developed was an idea of succession from earlier to later. According to him, all nature—and society as well—was connected and united in a cosmic harmony.

"The power which thinks and operates in me is in its nature a power as eternal as that which holds together suns and stars," Herder declared. Although the philosopher was willing to recognize that there were earlier and later species, he wasn't ready to acknowledge that those higher, more complex animals that did appear later might have been descended from earlier and lower ones. So, we can't say that Herder's views foreshadowed Darwin's except superficially. In Darwin's theory, organic evolution requires that more complex forms of life appeared on the Earth later than simpler ones, that these later forms of life were descended from earlier ones, and that the descent of these later species from the earlier was a consequence of variation and natural selection.

Who Are They?

Johann Gottfried von Herder (1744–1803) was not only an influential German philosopher, but he also was a literary critic who inspired several writers of the German romantic school, including the literary giant Goethe.

Buffon: The Death of the Biblical Chronology

By the close of the eighteenth century, the philosophical ground had been cleared of absolute prejudice against the idea of development, and men were taking a new and more empirical look at the history of nature itself. The biblical chronology was dying on its feet. The more empirical discoveries that were made in the fields of astronomy, physics, chemistry, and natural history, the more coherent and consistent a picture of the past scientists could build.

It took them some time, though, before the discovery of a widespread geological order in the nature of the Earth's crust was recognized as evidence of a temporal order. That is to say, the rock formations hadn't been plunked down all at once, but they had been assembled gradually, layer (or strata) by layer. Active processes had been at work through which the rocks had come into existence.

But it took most of a century before geologists established what agencies had been involved in the processes of formation. They had to learn how to compare the ages of strata geologically distant from one another and build up a consistent history of the Earth's crust from the evidence of its present form and fossil content.

The Importance of Buffon's Theories

By 1750, scientists began to recognize that the Earth might be much older than had previously been thought, extending back in the past even before humans appeared. But how old the Earth might be was still unknown. The first attempt to estimate the age of the Earth was accomplished in the 1770s by a courageous naturalist named Georges Louis Leclerc, better known as Comte de Buffon:

> [In] natural history, one must dig through the archives of the world, extract ancient relics from the bowels of the earth, gather together their fragments and assemble again in a single body of proofs all those indications of the physical changes which can carry us back to the different Ages of Nature.

Buffon had no doubt that the Earth had a long history or that this history could be reconstructed by human reason. Publicly, he ventured that the Earth might be at least 168,000 years old (considerably more than the biblical time scale would allot). Privately, he estimated that the Earth might be even older—half a million years old. (The truth is closer to four billion!) The Book of Genesis, he argued, was not intended for scientists, but for the unlearned. The days of creation couldn't have been actual "days." Instead, he said, we should look at these "days" as referring to geological epochs of indeterminate length.

He didn't have any intention of abandoning the biblical view of creation; what he hoped to do was somehow strike a compromise between science and religion. His attempt failed to satisfy anyone.

The Degenerating Species

Once Buffon had calculated that the Earth was far older than his contemporaries thought, he realized that there was enough time for species to have changed. But

Who Are They?

George Louis Leclerc, Comte de Buffon (1707–1788), was a French naturalist and the author of *Natural History*, a 36-volume work, that offered the first naturalistic account of life on Earth that incorporated such subjects as minerals, botany, and zoology.

there was nothing to be gained by trying to classify them—you'd be wasting your time. Why? Because of those confusing intermediate creatures that couldn't fit into any easy category: In general, the relationship between species is one of those profound mysteries of nature which man can only investigate by experiments which must be prolonged as they will be difficult."

He ridiculed the possibility that "even man, the ape, the quadrupeds and all animals might be regarded as making only one family." However, he was willing to accept the idea that existing species developed from original forms that had emerged when the Earth was still hot. Compared to these original forms of life, though, the present-day forms, said Buffon, were "degenerated" and dwarfish. As long as the Earth was still hot, new life could be produced spontaneously. Once the Earth began to cool, though, large-scale spontaneous generation was no longer possible.

Although Buffon was mistaken in many ways—creatures are not produced by spontaneous generation, for example, and man, apes, and all other animals actually do come from "one family"—his views did represent a leap forward. That's because he was willing to allow a dimension of time into his theory. He admitted, first of all, that species can change—a revolutionary idea in itself. Second, he determined that this change occurs over time. A hot Earth produces a lot of "original" forms, but once cooling starts, the descendants begin to "degenerate" and shrink. What he's saying is that organisms can change in response to environmental changes. That, too, is a great advance in thinking. We're still a long way from Darwin, but we're getting closer to the idea of natural selection.

Let's take a look at Buffon's major ideas:

➤ The Earth's age is between 165,000 and 500,000 years old, far older than the traditional biblical timeline.

➤ Species do change over time in response to a changing environment.

➤ The original species were produced spontaneously when the Earth was young and still hot.

➤ As the Earth cooled, species "degenerated" and became smaller until it was no longer possible to generate spontaneously.

The Albinos from Senegal and the Problem of Inheritance

Buffon's theories about the origin of life depended on a belief in the spontaneous generation of higher animals from "organic molecules," as he called them. Yet even before 1700, scientists had concluded that all the reputed kinds of spontaneous generation—mice emerging from soiled shirts or maggots created out of putrid meat—were false. There was no such thing as spontaneous generation; living creatures were produced only by others of their own kind. Even if the question of spontaneous generation could be put to rest, however, that still left the problem of how living creatures were created from their parents, whether they were bugs, primates, or humans.

By the late eighteenth century, the most widely held theories explained reproduction in terms of preformed seeds or germs, each generation of which contained all the elements needed to create another being of the same kind. Without an understanding of genetics and embryology, however, the problem would never be solved—and that was still a long way off. Only after you knew how a species could reproduce itself would you then be in a position to say how a species retained its identity—why, in other words, the offspring of the parents are of the same species.

Yet one eighteenth-century German thinker named Louis Moreau de Maupertius intuitively grasped how the identity of a species might be maintained in the short run and then altered in the course of evolution over the long run. Maintained and *altered?* Isn't there some contradiction here? Hardly. Mice produce other mice, and humans produce other humans. But over time—and we're talking about a lot of time—alterations creep in because of genetic mutations. As we'll see, new species can then be created or become extinct.

Let's get back to Maupertius. What makes him so special? A brilliant and versatile scientist, he was in charge of the Academy of Sciences in Berlin, under Frederick the Great, who ascended the throne in 1740. Maupertius became interested in the mechanism for inheritance when he learned of the arrival in Paris of two albino Africans. (Albinism is a genetic abnormality that depletes the skin of pigmentation.) Albinism, he found out, tended to run in certain families in Senegal. That suggested to him that such a peculiar trait must be inherited and that it should be possible to trace how it was passed on from one generation to the next.

By coincidence, Maupertius later encountered a surgeon on a train with an extra finger on one hand and an extra toe on his foot. The surgeon told the inquisitive German that this anomaly had turned up throughout his family tree—his mother and maternal grandmother had it, and so did three of his seven brothers and sisters and two of his six children. Maupertius realized that chance alone couldn't account for why such an anomaly recurred so frequently in a single family. His studies led him to the conclusion that in the process of sexual reproduction, both father and mother contributed a pair of "particles." Once united, each pair played a part in determining the development of the offspring. Some of these particles were dominant, making them more likely to occur in offspring. Occasionally an abnormal particle would reappear, and, if it were dominant, it could express itself in an anomaly such as albinism or extra digits. (Today, we would call these abnormal "particles" genetic mutations.)

However, Maupertius said, you didn't need to wait for chance to intervene to produce organisms with specific traits, whether good or bad. Farmers, herders, and dog lovers had long been breeding animals to ensure that the dominant traits would be passed along to the offspring by artificial selection. Using this method, men were creating "races of dogs, pigeons, canaries, which did not at all exist in Nature before." These custom-tailored races, Maupertius said, started out "with only fortuitous individuals," but "art and repeated generation have made species of them."

How these fortuitous changes came about, though, Maupertius wasn't sure: "Although I suppose here that the basis of all these variations is to be found in the seminal fluids themselves, I do not exclude the influence that climate and foods might have." All the same, he understood, as his contemporaries did not, that there was a direct connection between the problem of genetic inheritance and the problem of the species. "Could one not explain by that means (the fortuitous appearance of mutant 'particles') how from two individuals alone the multiplication of the most dissimilar species could have followed?" Perhaps, he suggested, this was true:

> The elementary particles fail to retain the order they possessed in the father and mother animals; each degree of error would have produced a new species; and by reason of repeated deviations would have arrived at the infinite diversity of animals that we see today; which will perhaps still increase with time ....

Maupertius was on to something crucial—errors in the form of genetic mutations do creep into the normal mechanisms of reproduction and inheritance. Over time, their cumulative effects can result in the creation of new species. Even though distinguished scientists such as Buffon knew about and discussed Maupertius's theories, the German was too far ahead of his time. And so his ideas languished for centuries because no one had the knowledge, technical expertise, or imagination to pick up where he had left off and refine or advance them.

The geological discoveries that led to the recognition that the Earth's past extended back into the mists of time (though how much time, no one was really sure) set in motion a cascade of zoological discoveries in the eighteenth and early nineteenth centuries. This is no surprise because the evidence that was being uncovered about the origin and development of Earth by geologists inevitably raised questions about the origin and development of life.

Before 1815, naturalists didn't think to ask about the development of species in time, because they were preoccupied by such questions about the existence of present-day species and how or whether to classify them, not how they originated or might have changed. Many naturalists, as we've seen, went as far as to argue that the classification of living things into distinct groups was a pretty arbitrary business anyway—a species, in their view, was an intellectual fiction. "Nature knows nothing but individuals," as one naturalist declared. So if you accepted the idea that there was no such thing as a definite species, there was no reason to ask how a species could either "originate" or "evolve."

There was the related problem of how species could retain their identity generation after generation and yet, over time, undergo alterations that might, if the alterations were significant enough, result in the creation of another species.

The Least You Need to Know

➤ Throughout much of the seventeenth and eighteenth centuries, doubt still existed about whether distinct species existed or whether they could be properly classified.

➤ Until the nineteenth century, it was not thought that organisms could develop over time, nor that "time," as measured from the creation of the Earth, encompassed more than 4,000 or 5,000 years.

➤ Without meaning to, eighteenth-century geologists uncovered evidence that showed that frequent catastrophes had occurred over the course of many years, which could not be accounted for by the biblical story of the Flood.

➤ Fossils unearthed by geologists revealed the existence of ancient ancestors of current species, as well as species that no longer existed.

➤ Some naturalists and philosophers, notably Comte de Buffon, recognized that the world was far older than the biblical timeline and that some form of change did occur in nature over time.

➤ An eighteenth-century German scientist, Louis Maupertius, recognized that anomalies—such as albinism or extra digits on hands and toes—could recur very frequently in certain families, and offered a surprisingly modern theory to explain how traits could be inherited.

Chapter 5

The Love Life of Plants

In This Chapter

➤ Carl Linnaeus: the great taxonomist

➤ The hidden life of plants

➤ Linnaean's hierarchy: from the simplest to the most complex forms of life

➤ Lamarck's view of natural selection: a blunder of historic proportions

➤ Erasmus Darwin: plants and poetry

How Earth reached its present stage of organization—geologically, chemically, and organically— continued to baffle and excite scientists of the eighteenth century. Yet, at the same time, even some of the most important naturalists of the day remained unwilling to surrender their belief in the story of creation as recounted by Genesis.

Although naturalists—most notably John Ray—had attempted a classification system of species, particularly plants, no one had successfully devised a system that was easy to learn and simple to use—and had the added advantage of making sense. It was a Swedish naturalist, Carl Linnaeus, who successfully achieved both these goals. His taxonomic system, while flawed in many ways, gave scientists a handy nomenclature to refer to species—until Linnaeus, three or four Latin names might have been used to designate the same plant. At the same time, Linnaeus organized species into a hierarchical scale, beginning with the lowest forms of life and culminating at the highest and most complex. Linnaeus's system still remains the basis for the standard of taxonomy used by scientists today.

However, Linnaeus was not so successful in recognizing how natural variation occurs over time. In an attempt to determine how such variation takes place, a French aristocrat with a passion for invertebrates—Jean-Baptiste Lamarck—conceived of a theory of natural selection that courageously tackled the problem but then took an unfortunate turn in the wrong direction. What Lamarck had failed to see—and what Darwin subsequently understood—was that evolution affects not individuals, but whole populations.

Carl Linnaeus: The Ultimate Classifier

The age of discovery, which had begun in the fifteenth century with the voyages of Columbus, Magellan, da Gama, and others, was reaching its apogee in the eighteenth century. For scientists, it was an exhilarating, if still unsettling, period.

A Natural Selection

Invertebrates are defined as organisms that lack a vertebrate or spinal cord. This includes the majority of creatures in the animal kingdom.

Explorers were constantly turning up new species of plants and animals to add to those already known. That led to more complications when it came to classifying them as a species. By the middle of the eighteenth century, two men—Buffon in France and Linnaeus, then a Swedish botanist—dominated the study of the natural order. Buffon, whom we discussed in the last chapter, didn't want anything to have to do with any classification system. Linnaeus, on the other hand, became the greatest classifier or taxonomist of all. (Today, he is referred to as the first of the modern "systematists.")

Why is classifying living creatures so important? Well, if you don't know whether a species is a mammal or a fish—dolphins and whales are mammals even though they inhabit the sea—it's impossible to begin to see how they are related to other species that exist at present or to those that existed in the past.

Like many great scientists, Linnaeus was an early disappointment to his parents, who had hoped that he would become a priest (although they took some consolation from his decision to study medicine at the University of Lund, Sweden). But Linnaeus's true love was plants. He came by his interest in botany naturally: His father was an avid gardener (although he made a living as a Lutheran pastor.)

What's in a Name?

Until Linnaeus appeared on the scene, species were often given long, complicated Latin names. What's more, there was no standard nomenclature to designate species, which meant that a scientist might find two or three names for the same species without realizing it. The common wild briar rose, for instance, was known as *Rosa sylvestris inodora seu canina,* but also as *Rosa sylvestris alba cum rubore, folio glabro.* It was confusing enough trying to classify species. And with explorers returning from voyages to the Americas, Asia, and Africa, with cargoes of plants and animals previously unknown in Europe, the need for a reliable naming system became much more urgent. One of Linnaeus's major contributions was to simplify the whole enterprise by designating one Latin name to indicate the *genus* and one as a "shorthand" name for the species.

The two names—designating the genus and the species—are known as the *binomial* (which means "two names.") For instance, Linnaeus renamed the briar rose *Rosa canina,* simplifying life greatly for future botanists. The advantages of this system were so obvious that the binomial system was soon adopted as the standard for naming species. The oldest plant names accepted as valid today owe their designation to Linnaeus's *Species Plantarum,* published in 1753. Linnaeus can also take credit for the oldest animal names, which first appeared in his *Systema Naturae* (1758).

What Does It Mean?

In biology, **genus** is a category that includes a group of species closely related in structure and evolutionary origin. Genus falls below family (or subfamily) in the classification of organisms.

Plants with Open and Hidden Marriages

Linnaeus's plant taxonomy was based on one thing: the reproductive organs—specifically, their number and their arrangement. Therefore, a plant's *class* was determined by its stamens (male organs), and its *order* was determined by its pistils (female organs). This method left something to be desired because many of the resulting categories seemed unnatural. The same categories could include plants with separate male and female "flowers" on the same plant, as well as multiple male organs attached to one common base.

Linnaeus's system also lumped in conifers such as pines, firs, and cypresses with a few true flowering plants, such as the castor bean. And then there were some "plants" that had no obvious sex organs at all. These were designated as belonging to the colorfully named Class Cryptogamia, or "plants with a hidden marriage." In this rather ambiguous class, we find the algae, lichens, fungi, mosses, and ferns, among others.

What Does It Mean?

A **class** refers to organisms with certain characteristics in common: Apes, monkeys and mice are all in the class Mammalia. Members of an **order** share more characteristics than members of a class: Dogs and raccoons belong to the class Carnivora because they are flesh-eaters.

Linnaeus acknowledged that his system amounted to an "artificial classification," not a natural one, which would take into account all the similarities and differences between organisms. Yet, like many naturalists of his day, Linnaeus placed such great importance on plant sexual reproduction (which had only recently been rediscovered) that he was willing to live with the inconsistencies and gaps. Linnaeus even went further and drew some unusual parallels between the sexual habits of plant life and humans:

> The flowers' leaves … serve as bridal beds which the Creator has so gloriously arranged, adorned with such noble bed curtains, and perfumed with so many soft scents that the bridegroom with his bride might there celebrate their nuptials with so much the greater solemnity.

Not surprisingly, basing plant classification on the sexual organs of plants proved quite controversial. One critic, a botanist by the name of Johann Siegesbeck, was so scandalized by the overt sexual references that he called the classification scheme "loathsome harlotry." (Linnaeus had his revenge, however, when he named a small European weed *Siegesbeckia*.) More importantly, though, the scheme, although undeniably easy to learn and useful, also produced results that didn't correspond with reality.

What Does It Mean?

Morphology refers to the study of structure or systems of a body—muscles or bones.

Subsequent systems of classification followed the practice of naturalist John Ray (who we discussed in Chapter 4, "Earth Gets a History"), who relied on *morphological* or structural evidence from all parts of the organism in all stages of its development. This represented a shift from Linnaeus's emphasis on one structure (reproductive organs) and one stage of life (sexual maturity).

The Linnean system is nonetheless a triumph for two reasons: It introduced a convenient (and standardized) method of naming species (the binomial nomenclature), and it provided future scientists a method of classifying organisms on a hierarchical basis (from simplest to most complex). His system has survived, in modified form, for more than 200 years, and his writings have been studied by every generation of naturalists, including Charles Darwin. The search for a "natural system" of classification hasn't come to an end, though. Today systematists are trying to develop classification systems on the evolutionary relationships of major categories of organisms based on common design or origin.

Why Is Linnaeus Important to Evolution?

Until Linnaeus, eighteenth-century naturalists had been testing Ray's system of classification to discover whether his ideas about plants had an application to all other forms of life. Was it possible, they wondered, to distinguish each kind of living crea-ture by describing some combination of characteristics? This creature has traits A, B, and C: Therefore, it must be X, not Y.

The problem that they confronted was that, to do this successfully, you need to have a grasp of the nature of reproduction—why did X parents produce X offspring and Y parents produce Y offspring?—which they did not. So the mystery that had given such a headache to naturalists for generations still persisted: Did species change, or didn't they?

In the first years of study that Linnaeus devoted to plants, he accepted the idea that species were a distinct category. (Remember that many others, including his contem-porary Buffon, did not think that way at all.) On the other hand, Linnaeus believed, at least initially, that species were unchangeable. "The invariability of species is the condition for order," he wrote, referring to the natural order. Time had not added to their number. Yes, he admitted, environmental differences might have affected species on a local level, causing their height or the color of their feathers to change, but these changes were of little significance, being both temporary and transient. That was what he thought, until he came to study a plant called the toadflax.

The toadflax, a native to Europe with distinctive flower heads, developed from a simple but frequent genetic mutation. Although Linnaeus had no knowledge of genetics—no one at the time did—he was forced to recognize that the hybridization could produce plants that certainly gave every appearance that they were new forms of life. This caused Linnaeus to admit, "It is impossible to doubt that there are new species produced by hybrid generation ...."

The evidence caused Linnaeus to abandon the concept that species were fixed and in-variable. Instead, he acknowledged, it was possible that hybridization was responsible for producing some or even *most* species that appeared after the creation of life. But Linnaeus was willing to go only so far. The process of generating new species wasn't unlimited in his view, nor could it go on indefinitely into the future. He still clung to the story of Genesis. Although new species had emerged since creation, he said, the original species (*primae speciei*) responsible for them all came from the Garden of Eden and so were still part of God's plan.

By 1762, he put forth the argument that God had originally created the parents, not of each species, but of each genus or order. In other words, God had created the mod-els or molds for future creatures. It was only later that the different species had come into existence as variants of these original forms. Even if the species that now exist weren't present in the Garden of Eden, Linnaeus maintained, they had always *poten-tially* been present, which was, in his eyes, practically the same thing.

Linnaeus, like Darwin a century later, also observed that a struggle for survival is al-ways going on in nature. Nature, he said, is a "butcher's block" and a "war of all against all." But, once again, he sought to reconcile his religious beliefs with his sci-entific understanding. Struggle and competition were necessary, he said, if the bal-ance of nature were to be maintained—and this balance, too, was integral to the divine order. The idea that evolution could possibly be open-ended—meaning that it could keep going indefinitely or that it was not influenced by any divine plan or might not have any predetermined goal—would have shocked Linnaeus.

A Natural Selection

Linnaeus hoped to use his knowledge of botany to make the Swedish economy less dependent on foreign trade. To do this, he tried to introduce valuable plants, such as cacao, coffee, tea, bananas, rice, and mulberries into Sweden, thinking that they might adapt and flourish. His formidable knowledge of botany, however, was no match for Sweden's cold climate, and his experiment failed.

Nonetheless, by 1778, the year of his death, he no longer held that the possibilities of interbreeding and hybridization were quite so limited as he had once thought. Of course, that is still not the same as admitting that some species could become extinct or that other species could change into forms that would make them unrecognizable to their ancestors.

Buffon and Linnaeus: A Glance Backward

Even though they disagreed on many things, Lianneus and Buffon shared certain common assumptions:

➤ Present populations of living creatures owed their origin to parents as highly developed as themselves.

➤ However the original ancestral stock of parents developed (whether by God, as Linnaeus believed, or by spontaneous generation, as Buffon held), all subsequent species are derived from them.

➤ The present population of species developed from the ancestral stock by a process of variation due to environmental influences or interbreeding.

➤ It is not possible for extremely complex organisms to develop out of very primitive simple ones.

➤ Neither Buffon nor Linnaeus had more than a passing acquaintance with fossil evidence. That meant that neither naturalist could have reconstructed the progression of organic life over vast stretches of geological time.

If you compare the similarity between the zoological views of Linnaeus and those of Aristotle 2,000 years earlier, you might think little had changed. Linnaeus had come around to recognizing that species weren't fixed—an advance over Aristotle—but the mechanism for how species did change was something that Linnaeus had only barely grasped. He'd observed that breeding—hybridization—could produce variation among species. Buffon understood that environmental change—the temperature of the Earth, specifically—could influence how species changed over time, with more complex forms emerging spontaneously early in the Earth's prehistory and less developed species appearing later.

The stage was set for a naturalist to try to account for variation with environmental influences as well as by breeding. The result was one of the most audacious attempts to conceive of an evolutionary theory. It also happened to be one of the most disastrous.

Lamarck: Looking for Answers in All the Wrong Places

Jean Baptiste Pierre Antoine de Monet, the Chevalier de Lamarck, is better known today as Jean-Baptiste Lamarck. Lamarck and Lamarckism have come to be associated with a scientific approach that can best be described (even literally) by the cliché "barking up the wrong tree."

Born in northern France in 1744, Lamarck seems to have lived and labored under an ill-favored star from early on. Ironically, he had all the advantages of birth and education. The great Buffon was responsible for giving him his start. Yet he bungled practically every opportunity he was given.

At the age of 17, fleeing a Jesuit seminary, where he was supposed to be studying for the priesthood, Lamarck set out to seek his fame and fortune in Paris. Intuitively, he realized that Buffon hadn't gone far enough in connecting the problem of the origins of the species and the problem of geological change. So far, so good. Understanding how the Earth had acquired its present form was a key to understanding how species developed. And Lamarck was right to be skeptical about Buffon's theory—the spontaneous generation of the first life forms, which he called "organic molecules"—because the fossil evidence failed to show anything of the kind. Here, again, Lamarck was right on target.

Assuming that the Earth had begun life as a fiery, formless planet, how did it achieve such an elaborate physical, chemical, and organic organization? Lamarck decided to concentrate on how evolution led to the change in the Earth's chemical properties. It was a wildly ambitious objective that amounted to what one writer calls "a self-inflicted wound that would ruin his reputation for the rest of his life."

In the 1790s, Lamarck was appointed to the Invertebrate Division of the Natural History Museum in Paris. For the next 25 years, he plunged into the studies of invertebrate fossils, which led him to conclude that there was a "gradual perfecting of the organization" of living things over the eons. The older the rocks were, he observed, the simpler the organic forms were; later rocks were associated with more complex organic forms. This is what is known as the doctrine of *progression*. Again Lamarck was on to something. So where did he go wrong? What was the nature of his historic blunder?

Lamarck was familiar with Linnaeus's taxonomy, but he didn't think much of it. On the contrary, Lamarck discounted the Linnaean system as chaotic and not very useful because it categorized all animals below vertebrates into two classes: insects and worms. He decided that he could improve on Linnaeus. Where the Swedish naturalist was fixated on the reproductive organs of the plants he was trying to classify, Lamarck was far more expansive when it came to invertebrates, which he classified according to their respiratory organs and circulatory and nervous systems.

Like Linnaeus, though, Lamarck believed that all major groups of animals could be slotted into a hierarchy, a step-like progression that he described as a "veritable chain." The veritable chain represented a progression extending from the simplest polyp at one end of the scale to humans on the other. Because humans were obviously the most complex forms and occupied the top step (shades of the ladder of being), any species on lower steps must have gotten there as a result of "degradation." Yet Lamarck was shrewd enough to hypothesize that the simpler "degraded" forms at the lower end of the scale provided the material for the more complex forms above them.

The theory was nothing short of revolutionary. It suggested that there was an evolutionary process at work, with ever-higher forms developing from the lower forms. (This contrasted with the view that all the species had their place to begin with and had no connection with one another by descent.) This is what Lamarck had to say:

> Citizens go from the simplest to the most complex and you will have the true thread that connects all the productions of nature; you will have an accurate idea of her progression; you will be convinced that the simplest of living things have given rise to all the others.

His contemporaries weren't terribly impressed. A "new piece of madness" was how the eminent geologist George Cuvier greeted this theory. (We'll get to Cuvier shortly.)

That later animals might be descended from earlier and simpler ones, gradually increasing in complexity from one geologic epoch to another, convinced Lamarck that the later animals were related to the earlier ones by actual descent. This line of descent was illustrated in his work, *Zoological Philosophy* (1809), which attempted to demonstrate how organisms developed along various branches, all traced back to the hypothetical point that Lamarck called the "first beginnings of organization." The origins of life, in Lamarck's scheme, were vaguely due to "heat and humidity."

Here's where Lamarck got into trouble. He was writing in the 1820s, when too little was known about biochemistry to make much headway in figuring out how life might have begun. Heat and humidity, as an explanation, gets you only so far. How to account for the appearance and survival of new and more complex forms of life? Any number of mechanisms suggested themselves. Lamarck didn't seem to appreciate the fact that, once you accept the theory of descent, you then have to go about establishing how it actually works. He struggled to come up with a plausible answer.

> Species have only a limited or temporary constancy in their characters, and ... there is no species which is absolutely constant. Doubtless they will subsist unchanged in the places in which they inhabit so long as the circumstances which affect them do not change, and do not force them to change their habits of life.

Reread that sentence, and you'll see that it doesn't say very much at all. Yes, species change—unless they don't.

Sometimes, he maintained, animals develop new limbs or organs as a somewhat unconscious response in coping with the new modes of life in their environment. On other days he credited organic progression to the "universal force" responsible for all organization in nature:

> The more frequent and longer sustained use of an organ gradually strengthens this organ, develops it, enlarges it, and gives it a power proportionate to the length of time it has been used. On the other hand, the constant lack of use of an organ insensibly weakens it, causes it to deteriorate, progressively diminishes its faculties, and tends to make it waste away.

As an example, he pointed out, a dog that spends the winter in the open would acquire a thicker coat than a dog that had been kept indoors. The pampered dog, after all, wouldn't need a thick coat, unlike the mongrel fending for itself in the wild. Intuitively, this makes some sense. Organisms do adapt—at least to some extent—to local conditions.

Where Lamarck took a wrong turn, though, was in his assertion that once an organism acquired a new trait— bigger muscles, say, or a thicker coat—that organism would then be capable of passing that trait to its offspring. It was this same process, according to Lamarck, that had transformed the simple organism into a complex one. Of all the examples of this supposed mechanism of inheritance, the one most widely cited by critics of Lamarck involves the giraffe. This is how Lamarck put it:

> We know that this tallest of mammals living in arid localities, is obliged to browse on the foliage of trees. It has resulted from this habit, maintained over a long period of time, that in all individuals of the race the forelegs have become longer than the hinder ones, and that the neck is so elongated that it raises the head almost six meters (20 feet) in height.

What's Wrong with This Picture?

Recall what we said in the first chapter: Evolution affects not individuals (dogs or giraffes), but only populations. Lamarck's theory didn't distinguish between the changeable traits of an individual—eye color, hair, muscles, height, and other attributes affected by climate, food, and the environment—and the basic characteristics of a particular population or species. We now know what Lamarck did not—an individual's appearance is a consequence of his genes.

Appearance, however, isn't responsible for our genetic inheritance. That's not to say that the environment doesn't have an effect on an individual's makeup. Of course, it does. If a dog (or any other animal) possesses genes that make it more likely to develop thicker fur in harsh, cold climates, then that dog is more likely not only to survive but also to pass along those "fur"–related genes. So Lamarck was right: The environment does have an effect on individuals. But he was also very wrong: Environmental effects on individuals are not passed along to offspring.

Perhaps Lamarck realized that his theory wasn't convincing because he continued to propose alternative mechanisms of organic variation. Each new theory of his seemed less plausible than the last. Not that the climate was congenial to the views of Lamarck. Lamarck's fellow scientists weren't willing to recognize the idea of descent to begin with. Nor were they eager to embrace the idea that variation in species could occur over time. This was an era when even naturalists preferred to accept the concept that lower species had somehow "degenerated" from the higher ones rather than considering the possibility that the lower ones had contributed to the evolution of the higher ones.

Lamarck died in 1829 at the age of 85, blind, despised by his colleagues, and with his concept of the Natural Order thoroughly discredited. Yet Lamarck's contribution shouldn't be underestimated. Scientific progress is never a smooth path. Mistakes are inevitable, and scientists often learn as much, or more, from the mistakes of their predecessors as they do from their successes. In that respect, Lamarck is a heroic—and tragic—figure in the evolutionary saga.

A Natural Selection

Jean-Baptiste Lamarck died in 1829 without ever receiving much scientific recognition for his work, which included the seven-volume *Natural History of Animals Without Backbones,* or his ideas. It wasn't until the latter half of the nineteenth century that his theories enjoyed renewed attention.

Barking Up the Wrong Tree

Let's go back to the example of the giraffe to see where exactly Lamarck went astray. How did the giraffe get its characteristically long neck? Lamarck said that those giraffes that kept stretching their necks so that they could reach leaves on high branches ended up with longer necks and then left offspring that would develop equally long necks. This makes no more sense than to say that if you became a top-notch tennis player, your children will also be great tennis players.

However, if you have genes that give you physical dexterity, you may well turn into a very good tennis (or basketball or volleyball) player, and if your children inherit some of those same genes, they might also turn out to be athletic. The giraffe acquired its long neck, Darwin said, not by stretching day after day to reach those coveted leaves, but because some giraffes were born with longer necks than others. (As we know now, that's because some giraffes had mutated genes coding for longer necks.) The lucky long-necked giraffes were obviously in a better position to get the leaves and so were better able to survive. It followed, Darwin said, that the survivors would be the ones to leave offspring. (Those short-necked giraffes would have starved to death and thus would have left few progeny, which, having inherited short necks themselves, wouldn't fare any better.) The difference between Lamarck's theory and Darwin's is profound.

Erasmus Darwin: The Poet of Nature

Finally we come to a man who had links to both Linnaeus and Lamarck: Erasmus Darwin, Charles's grandfather. Erasmus was attracted to Linnaeus's work because, like the Swedish taxonomist, he was interested in the possibilities of developing new species from plant hybrids. Hoping to introduce Linnaeus to a larger public, Erasmus pushed to get the first English editions of his works into print. Erasmus, who lived from 1731 to 1802, also shared many of the same ideas that Lamarck advocated and that would later be soundly repudiated by his own grandson.

A true eccentric, a physician, philosopher, poet, botanist, and free-thinker, Erasmus was one of the leading intellectuals of his age. Erasmus believed that existing life forms had evolved gradually from earlier species—an idea that would be incorporated into Charles's work 60 years later. Erasmus was interested in learning how life could have evolved from a single common ancestor, forming what he called "one living filament." His assertion that all vegetables and animals now existing were originally derived from the smallest microscopic ones, formed by what he called "spontaneous vitality," explicitly excluded divine intervention as having had any influence on creation—and it also happens to be close to modern scientific theory. In his posthumously published *Temple of Nature* (1802), Erasmus applied his poetic talent to his botanic studies when he offered this description of how life originated:

> Organic life beneath the shoreless waves
> Was born and nurs'd in ocean's pearly caves;
> First forms minute, unseen by spheric glass,
> Move on the mud, or pierce the watery mass;
> These, as successive generations bloom,
> New powers acquire and larger limbs assume;
> Whence countless groups of vegetation spring,
> And breathing realms of fin and feet and wing.

Erasmus also anticipated his grandson by raising the possibility that competition and sexual selection could cause changes in species: "The final course of this contest

A Natural Selection

In his book *Phytologia* ("The Study of Plants"), Erasmus Darwin offered one of the earliest detailed description of photosynthesis and also describes for the first time the geological principles of the artesian well.

among males seems to be, that the strongest and most active animal should propagate the species which should thus be improved." And he was on solid ground when he said that cross-fertilization was a superior form of reproduction to self-fertilization because it encouraged diversity.

However, he ascribed evolutionary development to the organism's conscious adaptation to needs and environment, a notion that has no basis in fact. In Erasmus's *Zoönomia,* or *The Laws of Organic Life* (1794–1796), a popular two-volume work on animal life, he put forward ideas that echoed those of Jean Baptiste Lamarck. Where he seems to have gone off the deep end entirely is in his statement that, from studying insectivorous plants, he was convinced that vegetables had muscles, nerves, and a brain.

The Least You Need to Know

➤ Carl Linnaeus made two significant contributions to the systematizing of nature: He provided a convenient and consistent standard for naming species, and he offered a hierarchical system that allowed scientists to categorize species, ranging from the lowest and most simplest forms of life to the most complex.

➤ Only late in his life did Linnaeus recognize that variation can occur in nature and that new species have emerged since creation because of interbreeding and hybridization.

➤ In an effort to discover how variation occurred among species, the French scientist Jean-Baptiste Lamarck focused on the effect of environmental change on individuals.

➤ Lamarck's theory of natural selection proved misguided because traits—positive or negative—developed by an individual (such as stronger muscles) cannot be passed along to the individual's progeny.

➤ A naturalist like his famous grandson, Charles, Erasmus Darwin believed in an evolutionary process in which existing species developed from simpler forms of life, but he also fell prey to Lamarckian thinking in asserting that organisms are capable of conscious modification in response to environmental change.

Part 2

The Advent of Darwin

By the middle of the nineteenth century, scientists were mostly in agreement that evolutionary change of some kind did take place in nature, but what they still didn't know was how evolution operated. What, in other words, was the mechanism of evolution—the engine that kept the whole thing going? The most penetrating thinkers of the day repeatedly came to grief in their attempts to explain how evolution might work. But in science, even mistakes and false starts can point the way to a fruitful avenue of research. The foundation was laid for somebody to come along and build a theory of evolution that would explain why some species evolved and others became extinct.

Charles Darwin didn't set out to revolutionize science; in fact, his father, a physician, considered him a wastrel who preferred to hunt and spend time with his indolent friends rather than apply himself to any disciplined course of study. Yet fortunately for science, Darwin possessed a keen intellect and unbridled curiosity. He once described his mind as "a machine for grinding out general laws out of a large collection of facts." In Part 2, we look at some of the most significant facts that went into the machine and then see how that machine ground out the general laws of evolution that would change history and cause the elder Darwin to reassess his views of his son's prospects.

Form, Function, and Catastrophe

It's such a cliché that probably few people stop to seriously ponder what "form follows function" actually means. And, is it possible to turn the phrase around and say with a reasonable degree of accuracy that "function follows form"? Well, in a nutshell, these two viewpoints are at the heart of a debate waged by two early nineteenth-century naturalists, both of whom held prestigious positions at the Museum of Natural History in Paris: George Cuvier and Etienne Geoffroy St. Hilaire. What makes the debate—which, in one form or another, still goes on today—more than of passing interest is that it goes to the heart of the question about how evolution takes place.

Cuvier said something quite dramatic—a real break from earlier thinkers. He declared that, based on the fossil evidence, species do become extinct. How? They're swept off the face of the planet by geological catastrophes—he called them "revolutions." You'd be tempted to say: Here is a thinker with a decidedly modern view. And you'd be right—but only up to a point. Cuvier said something else: He contended that organisms do not undergo change. That can't be allowed, he stated, because organisms are a functional whole—tamper with one part, and you upset the exquisite balance of the whole. Modification, which is what evolution is about, therefore is out of the question.

Cuvier reached his preeminent position as a zoologist, working at the Museum of Natural History in Paris shortly after the French Revolution (a very perilous time, by the way) as a result of the assistance of another zoologist, Etienne Geoffroy St. Hilaire. (Under French nomenclature, he is referred to more simply as Geoffroy, not St. Hilaire.) The two men were collaborators on many projects. But in spite of their often close association, they sharply disagreed on the way in which nature attained its current state. For Geoffroy, form was the key, not function. In a daring imaginative leap, he went further and tried to show how the structures of living creatures could reveal a common ancestor—he called it an archetype—a view that earned Cuvier's scorn.

So how did this debate play out? Who won? Did either of them? We'll see that both men were correct—up to a point. And both men were wrong. Yet each, in his way, made significant contributions to the understanding of nature's mysterious processes, which set the stage for the groundbreaking work of Charles Darwin later in the century.

George Cuvier: Rescuing Geology from the Atheists

While Lamarck had started out with all the advantages—a titled name and a brilliant mentor to launch his career—George Cuvier was the perpetual outsider, a Protestant of Swiss ancestry, who had no particular connections. Yet whereas Lamarck foundered in his endeavors, Cuvier experienced nothing but success. Entering Paris in 1795, at the age of 29, he soon became lionized as one of the "new men" of the post-Revolutionary era. (The French Revolution had occurred just six years before.) He soon received an appointment as a professor of animal anatomy at the newly formed National Museum of Natural History, where he specialized in the study of fossil vertebrates: mammoths, mastodons, and extinct forms of the rhinoceros, hippopotamus, deer, and crocodile.

A Natural Selection

By the end of his life, George Cuvier was so familiar with the skeletal structures of the vertebrates that he boasted that he only had to be shown a single fossilized bone of an unknown species to describe the whole animal.

Cuvier Claims Lamarck Can't Adapt

In Cuvier's eyes, Lamarck had concentrated on the highest forms of life in any given era—"the spear point of evolution"—to the exclusion of the simpler forms. Besides, Cuvier argued, Lamarck had laid out a continuous progression—lower to higher forms of life—without taking into consideration that new species (lower, higher, or in between) might emerge or disappear in different geological epochs. It wasn't enough to explain how the lowest forms of life had originated or by what mechanisms new and more complex organisms were produced in each epoch, declared Cuvier. If a theory was going to hold water, you also had to explain two things: first, how living species of every kind are so perfectly adapted to their present environment, and, second, why similar species in the past have disappeared altogether. The relationship between adaptation and extinction was one of the crucial questions that Darwin later grappled with.

Fossils had long been accepted as the remains of once-living organisms. Some scientists before Cuvier, including Buffon, believed that many fossils represented life forms that no longer existed. For instance, Buffon wrote prophetically:

> We have monuments taken from the bosom of the Earth, especially from the bottom of coal and slate mines, that demonstrate to us that some of the fish and plants that these materials contain do not belong to species currently existing.

His view was sharply disputed by other scientists of the day who simply could not believe that God, having created all life, would allow any of his creations to be swept off the Earth.

So what other explanation could there be for fossils if they weren't from ancient species? Well, some scientists contended that fossils were actually the remains of living species. Fossils of mammoths found in Italy, they said, were the remains of the elephants that Hannibal brought with him when he set out to invade Rome. Other theories were proposed, including one that proposed that fossils were the remains of unusual organisms that still flourished in parts of the world that hadn't been explored yet. Even Thomas Jefferson thought it possible that mammoths might turn up somewhere in the American West. Learned folks could speculate all they cared to, but only until evidence was collected and thoroughly examined could the problem be solved.

Cuvier certainly recognized the need to gather more information, as he stated in a 1796 talk that he gave about finds of fossil elephants: "The studies of elephant bones published up until now contain so little detail that even today a scientist cannot say whether they belong to one or another of our living species." Yet eventually Cuvier was able to conclude that many fossils did represent evidence of extinct forms of life.

Catastrophic Evidence

How did species become extinct? Cuvier thought he had the answer:

> Life on this earth has ... frequently been afflicted by these terrible occurrences. Numberless living things have been the victims of these catastrophes; those living on dry land having been swallowed up by floods, others whose home was in the waters being left high and dry when the sea-beds were suddenly lifted up. Whole races have been wiped out for ever, leaving no record in the world but those few relics which even the naturalists can barely recognize. (The translation from the French here is a little loose: Cuvier didn't use the word *catastrophe*, preferring *revolution*. *Catastrophe* had supernatural overtones that he didn't like.)

Examining the geological evidence, Cuvier was struck by a phenomenon that he termed *discontinuities:* Organic fossils embedded in a lower strata or layer of rock would disappear in the strata above, replaced by fossils of different species or, in some cases, no fossils whatsoever. How could these discontinuities have occurred? Only convulsive upheavals of the Earth's crust, Cuvier argued, could explain why the fossil evidence didn't exhibit a clear line of descent from lower to higher forms. Examining the geological evidence, a species of fish, for example, that once had swum in a pond that had receded, would disappear and leave behind no progeny. Catastrophes, though, were a fairly rare occurrence in the Earth's history, he said, and were bracketed by long periods of relative tranquility.

Each new geological epoch, Cuvier said, would produce animals and plants that were new and not related to species of former epochs. That implied that each newly emerged species would also readily adapt to its new environment. If Cuvier was right (and, as you'll see, he wasn't), where did this leave the human race? Cuvier had an answer to this question, too. Humans were not more than a few thousand years old, capable of tracing their ancestry back only to the last two geological eras. Just as the Bible said, humans managed to survive the Flood—the last documented catastrophe— only because of the intervention of divine providence.

Where Cuvier departed from the traditional biblical story, was in his belief that the Earth itself was immensely old—far older than the 5,000 years calculated by early theologians—and, for most of its history, had experienced conditions more or less like those seen at present. Nonetheless, from time to time the Earth was subject to catastrophes, such as the biblical Flood, each of which had wiped out a number of species. These "revolutions" were all due to natural causes, Cuvier said, and it was up to geologists to consider their causes and effects.

Of one thing Cuvier was quite convinced: The Flood was responsible for carving the Earth into its present shape. In fact, geological studies would make no sense "without recourse to a deluge exerting its ravages at a period not more ancient than that announced in the Book of Genesis." But from the geological evidence, it was clear to Cuvier that other catastrophes must have taken place as well. Apart from the Flood, however, the zoologist didn't specifically relate any of these additional "revolutions" with biblical or historical events.

As one of the most influential intellects of his time, Cuvier's views brought renewed interest in mass extinctions and geological catastrophes as having caused at least some of the greatest episodes of change in life forms on Earth. Cuvier's recognition that extinction is a major force in nature represents a significant scientific achievement.

Although he was perceptive enough to understand that species do become extinct and that new species emerge in different geological epochs, Cuvier did not believe in organic evolution. Organisms, he said, were integrated wholes, where every part's form and function was in harmony with every other part's form and function. Any change in one part would destroy the delicate balance. By the same token, any one part of an organism, no matter how small, exhibited signs of the whole being. That is to say, you would need only a few fragmentary remains, in theory, to reconstruct the whole creature from which they came. Because of these ideas, Cuvier became known as the founder of the "functionalist school."

Any similarities that existed between organisms of different species, said Cuvier, could be accounted for by functions that their parts had in common. "If there are resemblances between the organs of fishes and those of the other vertebrate classes, it is only insofar as there are resemblances between their functions," he wrote. He rejected any notion that species might have a shared ancestry.

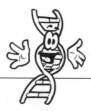

A Natural Selection

Cuvier remained as professor of the Museum of Natural History after Napoleon came to power, and he served in several government positions under Napoleon's regime. He then continued to hold positions under three successive kings of France, achieving the incredible feat of serving under three different, opposing French governments—revolutionary, Napoleonic, and monarchic. He died in bed.

In Cuvier's theory, function determines form; form does not determine function. Modifying even one part would impair the function of the whole. Cuvier opposed evolution because, by definition, organic evolution required modification. He maintained that any such changes would so impact an organism's anatomy that it would make it impossible for that organism to survive. When Cuvier had the opportunity to examine mummified cats and ibises from ancient tombs brought back by Napoleon's army from Egypt, he observed that they were the same as cats and ibises currently on Earth. This similarity of forms he took as compelling evidence that life forms do not change over time.

Cuvier's Dilemma

Cuvier's theory had a problem—a problem that was the reverse of Lamarck's, but one that was just as troublesome. Let's take the case of the mammoth and the elephant. The two bear a certain resemblance to each other, but the former is extinct and the latter is still among us. (Although, because of poaching and environmental depredation, the elephants may be in danger of extinction as well, outside of zoos.)

According to Lamarck's theory of descent, mammoths and elephants would be genealogically related—elephants descended from mammoths, with their changes explained by adaptation to changing environmental conditions. That would be why they are so similar to each other. Lamarck's failure was his inability to explain how the changes occurring in successive generations of woolly mammoths had finally culminated in the emergence of an elephant.

Cuvier's problem was the reverse. Assuming that the Earth has been visited by frequent "revolutions" that wipe out species and that fresh species are created in the geological epochs that follow, the question is, how do you get from here to there—from mammoth to elephant? If the mammoth's disappearance is explained by a catastrophe of some kind and then, in a new geological era, an elephant makes its appearance on the scene, how do you account for the resemblance then? If there is no line of descent, you have to come up with some alternative explanation.

Cuvier was mindful of the problem. In every age, he said, the fossil record reveals species belonging to four recognizable "branches"—the vertebrates, the mollusks, articulated creatures such as insects, and an assortment of creatures that he described as "radially symmetrical" or simply as *radiata*. Think of these branches as models or master plans. Any species that were brought into existence, adapted for life in the new geological environment in the aftermath of catastrophe, had to be designed to conform to one of these four basic structural branches. Having conceived these four branches, Cuvier used them to organize a sequence of classes, ranking them from lowest to highest. The orders in each class could be similarly ranked, and so on, down to the species level, with *Homo sapiens* inevitably sitting at the top of the scale of life.

Cuvier's theory held considerable popular appeal because it appeared to support the idea of divine creation. Scientifically, though, the theory left much to be desired by avoiding a question that both Buffon and Lamarck had struggled to answer: How did the

What Does It Mean?

Radiata, or radially symmetrical, animals have bodies that are symmetrical around a central point. Jellyfish, for instance, are radiata.

first generation of life get here? Lamarck had accepted spontaneous generation, at least in the case of the lower forms of life, while Buffon theorized that higher animals were the product of some kind of "organic molecules." Cuvier, though, preferred to duck the problem. How each new flora and fauna had come into existence was a question that he left deliberately vague.

What survived Cuvier wasn't his views on organic life as much as his geological theory of catastrophes, which its proponents seized as a vehicle for "rescuing geology from the atheists." Although geological change—even dramatic change—was allowed in his scheme, organic change was not. Species flourished or were wiped out, but they did not evolve. Cuvier had attempted to demonstrate that you could still credit God for bringing life into being and accept Cuvier's theory of new species with good conscience.

Cuvier died in 1833—four years after Lamarck—leaving behind 12 large volumes devoted to the fossil bones of quadrupeds. Cuvier's ideas exerted a great deal of influence for many years after his death, especially in France. He had successfully fought a delaying action, at least in the short run, against evolution by constructing an ingenious theory based to a large extent on unsound science.

So what can we say about Cuvier? Here are his basic ideas:

➤ The age of the Earth is immensely old.

➤ Geological "revolutions" or catastrophes—in addition to the biblical Flood—have occurred several times over the lifetime of the Earth.

➤ These "revolutions" have destroyed many species whose presence is recorded in fossilized remains; each new geological era produces fresh species, fashioned by the catastrophe and adapted for the new environment.

➤ Each organism is an integrated whole. Each part of an organism has a particular form and function that is in balance with the forms and function of other parts. Modification of one part will impair the function of the whole being. This is what is meant by the functionalist school of thought.

➤ Because modification of an organism impairs its function, organic change or evolution is impossible because the organism would not be able to survive.

➤ The resemblance of existing species, such as elephants, to extinct species, such as mammoths, is due to the similarity of the parts (and their functions), not because they shared a common ancestor.

Buckland Joins the Rearguard Action

For the first few decades of the nineteenth century, some sort of reconciliation between the geological evidence and the scriptural account of creation still seemed possible. Attempting to forge such a compromise, Cuvier had an ally across the English

Channel in a British zoologist and Anglican minister named William Buckland (1784–1856).

Buckland was also convinced that the fossil evidence had to be taken seriously. The Earth, he said, must have developed over a long period of time that extended back beyond the events chronicled in the Bible, including the Flood. Like Cuvier, he was willing to acknowledge that catastrophes accounted for much of the geological change over the millennia, and he also credited more routine processes—the effects of erosion and land elevation and flooding—with leaving a significant impression on the Earth's surface. At no point does he seem to have thought that these concessions to science would call into question the biblical account of creation (whenever in the calendar it turned out to fall).

Louis Agassiz: Wisdom, Greatness, and Providence

Cuvier may have been mistaken in his insistence that geological catastrophe was the key to understanding the natural order or that organic life could be divided into four branches and categorized from the lower form of life to the highest. But he inspired the work of several scientists, all of whom made significant contributions of their own—even if they all ended up on the wrong side of history.

The best-known disciple of Cuvier is considered one of the "founding fathers" of the modern American scientific tradition. Louis Agassiz, the son of a minister, was born on May 28, 1807, in Switzerland. In 1831, he went to Paris to study comparative anatomy under Cuvier. So impressed was Cuvier with Agassiz's work on fossil fish that he gave the young man his own notes and drawings for a work that he had planned to write on the same subject. Theirs was a brief association, lasting only six months before Cuvier died. Nonetheless, Agassiz considered himself Cuvier's intellectual heir. Assuming the role of disciple in earnest, Agassiz tirelessly defended Cuvier's views, stressing the importance of geological *catastrophism* and his classification system of organic life.

What Does It Mean?

Catastrophism is a theory, advocated by George Cuvier, that explains the present state of the Earth's surface as having come about through a series of violent geological cataclysms over vast periods of time.

Following Cuvier's lead, Agassiz asserted that the Earth had been periodically ravaged by global catastrophes, which set the stage for new species of animals and plants. In a novel twist, though, Agassiz abandoned the Flood as the primary instrument of geological change, replacing it with glaciers—he called them "God's great plough"—which he thought had been formed instantaneously throughout the world at the same time. Agassiz refined Cuvier's classification scheme as well. Cuvier's system was originally based on four branches: vertebrates, insects, worms, and a group that he lumped together as "radiata." The "lowest" forms of life, Agassiz said, were those that were found in the earliest strata of rock.

Agassiz, a contemporary of Darwin, became a fervent opponent of evolution. Far from acknowledging the role of natural selection—which is the basic mechanism for evolution and allows no role for any deity—Agassiz saw in all of nature evidence of a divine plan. He even went so far as to define a species as "a thought of God." His approach could be called an argument from design, which, even today, is used to counter Darwin's theory of evolution:

> The combination in time and space of all these thoughtful conceptions (species) exhibits not only thought, it shows also premeditation, power, wisdom, greatness, prescience, omniscience, [and] providence. In one word, all these facts in their natural connection proclaim aloud the One God, whom man may know, adore, and love; and Natural History must in good time become the analysis of the thoughts of the Creator of the Universe ...."

Agassiz's theory represents the last great expression of the old school of natural theology, dating back almost two centuries to John Ray, the English naturalist who also sought to find a divine plan in nature.

Even though he was mistaken on many counts, Agassiz did understand that no classification system worth its salt was going to be very useful if it didn't take into account a variety of factors, including embryology, the environment, and paleontology. Only by drawing on evidence from several different sources could you show how organisms are related to one another. If you say that a turtle is a lower form of life than a monkey, for instance, you have established a relationship between the two creatures, to be sure, but you haven't said anything about how or why a primate such as a monkey is superior to a turtle. However, if you provide information about their respective brains and functional capacities, someone would be in a better position to judge the accuracy of your statement.

Whatever the flaws in his theories were, Agassiz was insightful enough to recognize that several factors had to be considered in determining where exactly in the natural order an organism belonged. He wrote:

> Classification seems to me to rest upon too narrow a foundation when it is chiefly based on structure. Animals are linked together as closely by their mode of development, by their relative standing in their respective classes, by the order in which they have made their appearance upon earth, by their geographical distribution, and generally by their connection with the world in which they live, as by their anatomy. All these relations should, therefore, be fully expressed in a natural classification; and though structure furnishes the most direct indication of some of these relations ... other considerations should not be neglected which may complete our insight into the general plan of creation.

Geoffroy St. Hilaire and the Search for a Single Ancestor

"What is the essence of life—organization or activity?" asked the science historian E.R. Russell. That was at the heart of the great biological controversy of the first half of the nineteenth century. We've seen how Cuvier insisted that organisms are integrated wholes and that function determines form. But was it possible that Cuvier had gotten it backward—that form determined function? That was the view of the eminent zoologist and Cuvier's one-time patron, Etienne Geoffroy St. Hilaire.

In his 1818 book *Philosophie Anatomique,* Geoffroy asked the question: "Can the organization of vertebrate animals be referred to one uniform type?" Yes, Geoffroy answered. All vertebrates were simply variants of a single archetype, a primal form. If you accept the idea that all vertebrates derive from a common archetype—a kind of generic vertebrate that comes in several different varieties—then you could account for why *vestigial organs* exist that serve no functional purpose nowadays. The implication is that the vestigial organ had once served some function, but, over time, the organ was no longer needed. Because, according to Geoffroy, form trumps function, then it follows that that organ would remain, however functionless. Geoffroy's viewpoint is completely opposed to Cuvier's functionalist theories. "Animals have no habits but those that result from the structure of their organs," Geoffroy wrote.

In spite of their differences, Geoffroy and Cuvier enjoyed a close professional association. Geoffroy sponsored Cuvier's appointment to the National Museum of Natural History in Paris and collaborated with him on several projects subsequently. Geoffroy accompanied Napoleon's troops as a scientist when they invaded Egypt in 1798; on his return to Paris, he brought with him many animal specimens that he had discovered in his travels, including mummified cats and birds, which Cuvier used to prove that evolution had not occurred.

What Does It Mean?

A **vestigial organ** is one that serves little or no purpose, such as the appendix or tailbone in humans. Some flightless insects have tiny wings that are vestigial and certain snakes have vestigial leg bones, suggesting that early in evolution they once had a function that they have subsequently lost.

The Change Remains the Same

Geoffroy devoted a great deal of time drawing up rules for deciding when structures in two different organisms were variants of the same type—in modern terminology, when they were *homologous.* He looked for connections between parts: Structures in different organisms, he determined, were the same if their parts were connected to each other in the same pattern. Here is how another nineteenth-century naturalist of note described Geoffroy's contribution to understanding the relationships of different species by taking into account the similarity of their patterns of connection:

> What can be more curious than that the hand of a man, formed for grasping, that of a mole for digging, the leg of the horse, the paddle of the porpoise, and the wing of the bat, should all be constructed on the same pattern, and should include the same bones, in the same relative positions? Geoffroy St. Hilaire has insisted strongly on the high importance of relative connexion (sic) in homologous organs: the parts may change to almost any extent in form and size, and yet they always remain connected together in the same order.

The author of these words is none other than Charles Darwin.

Who Are They?

Etienne Geoffroy St. Hilaire was born on April 15, 1772, outside Paris. While he was studying medicine and science in Paris, the Reign of Terror broke out and Geoffroy risked his life to save some of his teachers and colleagues from the guillotine. After he managed to save his own head, he later secured appointment as professor of vertebrate zoology at what was to become the National Museum of Natural History in Paris.

The significance of Geoffroy's concept lay in his understanding of *homologous structures*. The identification and definition of these structures provided Darwin and evolutionary biologists who followed him with an important tool for establishing evolutionary relationships. If you understood that a human arm, a bat's wing, and a whale's flipper were all structurally related, you could then reasonably use this information to reconstruct a line of descent that would ultimately take you back to a common ancestor.

Geoffroy was so enthralled by his concept, however, that he took his "unity of type" concept farther than the evidence available warranted. For example, he came up with a theory that the segmented

What Does It Mean?

Homologous structures are those that evolved from a common ancestor. A human arm, a bat's wing, and a whale's flipper, although different in appearance, all have the same underlying structure.

external skeleton and jointed legs of insects were equivalent to the internal vertebrae and ribs of vertebrates. Insects, he said, literally live inside their own vertebrae and walk on their ribs. He was so carried away with his idea of a common ancestor for all species that he was supposed to have said: "There is, philosophically speaking, only a single animal."

Ultimately Unsound

Like many of the scientists we've been considering, Geoffroy grasped some crucial core ideas concerning evolution. At the same time, however, he failed to develop a sound theory based on the evidence at his disposal. We cannot say that he is an evolutionary biologist in anything like the modern sense. His expertise was in the field of morphology—the study of form, not how forms evolved through history. The common ancestral structures that Geoffroy used to make his case were ultimately *analogous structures* rather than homologous structures, which could actually be traced through the ages to existing species. He was on to an important point when he focused on the idea of a common ancestor; it was just that he wasn't able to prove that it was anything other than a lucky guess on his part.

So which is it—form or function? Neither Cuvier nor Geoffroy were right on all counts—their theories were both badly flawed, as we've seen—but could we say that one was closer to the truth? Does the form of an animal determine its habits and behavior? That's Geoffroy's position. Or, was Cuvier closer to the mark when he asserted that the function determines what kind of form an organism will develop when it comes time to make its appearance on Earth? (Recall that Cuvier didn't believe that species could change over time; each new geological epoch, set in motion by a catastrophe that basically wiped the slate clean, introduced fresh species to supplant the extinct ones.) This debate actually continues in some form today between what are known as the "formalist" and "functionalist" schools.

What Does It Mean?

In contrast to a homologous structure, an **analogous structure** may often look similar, but its underlying structures are quite different. While the tailfin of a fish and a whale's flukes look similar and perform similar functions, for instance, their structures do not look the same at all.

In some respect, however, the debate is already resolved in a compromise. Both schools are right: Organic lineages do change with time, in response to changing environments (which is more along the lines of Geoffroy's thinking). On the other hand, their form constrains the functions that they can assume in response to these environmental changes. Polluted seas, for instance, make it harder for dolphins or whales to survive—an environmental change—but their form prevents them from opting to try life on land. (So, in that sense, Cuvier was correct.) Modern evolutionists would say that the ability of organisms to function in their environments is critical to their evolutionary fitness and that form is often altered to fit a particular function.

Cuvier and Geoffroy were like the blind men who try to describe an elephant by touch alone. One found a limb and another found the trunk, but neither one understood the true nature of the animal being examined.

The Least You Need to Know

➤ Stressing the importance of the functions of an organism's parts over the organism's form or structure, nineteenth-century zoologist George Cuvier theorized that because all working parts of an organism are in balance, any modification in any part would impair the function of the whole organism, and, therefore, organisms cannot evolve.

➤ Because of fossil evidence, however, Cuvier believed that organisms can become extinct as a result of geological "revolutions" or catastrophes—of which several had occurred in addition to the biblical deluge.

➤ In contrast to Cuvier, Etienne Geoffroy St. Hilaire declared that organisms do undergo change and that these changes (of structure) are in response to changes in the environment.

➤ By identifying and defining similar structures in different organisms called homologous structures, Geoffroy noted similarities in species as diverse as humans, whales, and bats, suggesting that at one time in the distant past, they had all shared a common ancestor.

➤ Formalism (Geoffroy's school) emphasized the importance of the form or structure of an organism over the function of any of its parts.

The Heroic Age of Geology

The history of evolution as a theory is one to which many different fields have made significant contributions—before and after Darwin. At various periods in history, one field predominates over another—botany, say, or, in the ancient world, philosophy— but now we've reached a point in our story where geology plays the dominant role.

The mid-1800s are known as the Heroic Age of Geology for good reason. This was a period when the great geological epochs were defined and when much exploration and fundamental research was carried out. And it was an age that produced geologists of the high caliber of William Smith and Charles Lyell.

During the first half of the nineteenth century, geology came into its own as a science. Geologists began to recognize that rock formations are composed of layers or strata, each with its own characteristic mineral composition, reflecting the time period in which it was formed. What's more, surveying rock formations in different parts of the Earth showed that the sequence of these strata was the same, no matter where you went. Now, geologists had a reliable method of studying the history of the Earth. But that was only a prelude to the most exciting discovery, which carried extraordinary implications for understanding the history of life on Earth.

Although fossils had long been known and collected, their significance mainly went unrecognized; they were not generally regarded as evidence of species that had inhabited the world in the distant past and, in many cases, had vanished forever. The same fossils, geologists discovered, appeared in the same strata of rock, regardless of their location. That finding meant that naturalists could tell in which order life had appeared on Earth.

These discoveries hardly resolved the problems that scientists had been wrestling with for years. Heated disputes broke out on whether the geological record was complete; geological changes over the years might have eliminated many important fossils. And what had happened to fossils of organisms that no longer seemed to exist? Maybe they were simply hiding out in parts of the world that had yet to be thoroughly explored. Even supposing that species had become extinct, how had that happened? As we've seen in the last chapter, Cuvier and his disciples held that catastrophes, including but not limited to the Flood, were responsible both for major geological changes on Earth and for the extinction and creation of species. Other scientists disagreed, saying that it wasn't necessary to resort to catastrophes to explain the current state of the Earth's surface and that, over time, even gradual changes, such as erosion or a succession of storms, could produce the same effects.

In this chapter, we'll focus on three great geologists—William Smith, Charles Lyell, and Adam Sedgwick. We'll see how the geological and fossil discoveries of the early nineteenth century—and the disagreements over their significance—played out and how they laid the groundwork for Darwin's theory of evolution.

William Smith: Digging into the Rocks

From the 1600s through the 1800s, a formal education wasn't necessary to become a scientist. Certain disciplines weren't even taught at most educational institutions. (Chemistry, for instance, wasn't considered a serious subject for study until late in the nineteenth century.) Certainly many of the most important pioneers in the story of evolution had no formal training in the field in which they would later excel. This was true of William Smith, a self-made craftsman who was born on March 23, 1769, in Oxfordshire, England.

Although Smith received only limited schooling while growing up, he taught himself geometry, surveying, and mapping, all skills that later came in handy for his future explorations. As a child, he took an interest in collecting fossils, which he easily found near his home. By the time he reached the age of 18, he found work as an assistant surveyor of canal routes, a job that allowed him to discover thousands of more fossils.

To do his job, Smith first needed to acquire detailed knowledge of the rocks through which the canals would be dug. What struck him as he examined the rocks was how the fossils found in a section of sedimentary rock invariably appeared in a certain order from the bottom to the top of the section. That finding alone might have been a mere curiosity had it not been for the fact that the same order of appearance showed up in other rock sections—and even in rock formations on the other side of England. As Smith described it, "each stratum contained organized fossils peculiar to itself, and might, in cases otherwise doubtful, be recognized and discriminated from others like it, but in a different part of the series, by examination of them."

Smith's words are a statement of the "principle of faunal succession":

➤ Sedimentary rocks of different ages will contain fossils of different species.

➤ Knowing the age of a fossil helps to date the rock in which it is found.

➤ The layers of sedimentary rocks in any given location contain fossils in a definite sequence.

➤ The same order can be found in rocks elsewhere, allowing the comparison of strata in different locations.

The principle of faunal succession that arose from Smith's observations still holds today, although it has undergone some modifications. (Some fossil species, for instance, aren't very useful for making comparisons because they may not be distributed over a wide area.) Smith's work launched a new science called *stratigraphy* (derived from the word *strata*), which transformed geology into a historical science. His work provided convincing evidence that the strata in all parts of the Earth's crust appeared in a single common sequence—Cambrian, Ordovician, Silurian, Devonian, and so on.

Moreover, this sequence of rock wasn't a geographical phenomenon, but it offered a glimpse into the history of Earth: The order of rocks also reflected the order in which the rocks had been laid down during the formation of the planet. The common order of the strata also conveniently corresponded to the sequence of fossil species that they contained. Using the fossil evidence as confirmation, geologists now had an additional means of assessing when the strata formed in addition to determining their mineral content.

A Natural Selection

Index fossils are formed from species that existed only for a short time; they are particularly valuable in determining a rock's age. Two rocks from different locations that contain the same index fossil must be approximately the same age.

What Does It Mean?

Stratigraphy is the study of the layers of the Earth's surface.

By the 1820s, largely as a result of Smith's work, geology was turning into a developed natural science. What made life for geologists easier, though, was making it harder for zoologists. Was it really possible that the Earth's population of living things was so dramatically different from one era to another? Or did these findings mean that some fossils were more durable than others? (In other words, what geologists were discovering in the fossil record wasn't all there was, so some important fossils had disappeared because of geologic change.) But supposing that species were changing or vanishing completely, as these fossils seemed to indicate, did this mean that a succession of new creatures was constantly emerging from the mists of time? That, as we've seen, was Cuvier's argument. On the other hand, perhaps organisms were actually changing their forms, which was Geoffroy's position.

Like Cuvier and many other early naturalists, however, Smith was not about to let the fossil evidence that he had scrupulously analyzed disturb his religious faith. He seemed to see in the fossils that he had unearthed evidence of a divine master plan. He wrote:

> If, in the pride of our present strength, we were disposed to forget our origin, our very speech betrays us, for we use the language which he (God) taught us in the infancy of our science. If we, by our united efforts, are chiseling the ornaments and slowly raising up the pinnacles of one of the temples of nature, it was he that gave the plan, and laid the foundations, and erected a portion of the solid walls, by the unassisted labour of his hands.

One Too Many Catastrophes

By the late 1820s, Europe's geology had been carefully surveyed and mapped, and a great deal had been learned about the terrain in other parts of the world as well. Geologists were now in a fairly good position to say that, given enough time, the present condition of the Earth's surface could be explained by geological processes alone.

But what processes, exactly? The prevailing theory was that it was due to catastrophes, as Cuvier had advocated. There was just one problem—well, actually two. In the first place, the geological record seemed to demand an excessive number of catastrophes—as many as 27, by some counts. The second problem was reconciling Genesis with the geological record, a problem that Cuvier, Buckland, and Smith struggled mightily to resolve. Did these catastrophes represent repeated interventions on the part of divine providence as the Flood had? The idea seemed ridiculous. So it was only a matter of time before the catastrophist compromise was destined to collapse. The person responsible for pulling down the theoretical scaffolding that Cuvier had spent his whole career erecting was a young Scottish geologist named Charles Lyell.

Lyell, along with William Smith, is considered one of the founders of stratigraphy. Among his other accomplishments, he developed a method for classifying strata, or layers, by studying fossils of marine life in the rocks of Western Europe. Marine beds closest to the surface—which were the most recent strata—contained many species of shell-bearing mollusks still in existence. However, when he excavated the deeper, older strata, he discovered increasingly fewer fossils of living species. He divided the rocks of this period into three epochs, based on decreasing percentages of modern species. The names that he applied—Eocene, Miocene, and Pliocene—are still used today.

Disdainful of catastrophism, Lyell punctured Cuvier's theory. First, Lyell said, the idea of discontinuities—those mysterious gaps in the fossil record in the rock strata—was based on a misconception. That was because the discontinuities were only local, Lyell said. Fossils might not appear in one strata in Leeds, for example, but that didn't mean that they wouldn't show up in the same strata in Bath:

> We often find, that where an interruption in the consecutive formation in one district is indicated by the sudden transition from one assemblage of fossil species to another, the chasm is filled up, in some other district, by other important groups of strata.

Then Lyell turned to the second part of Cuvier's catastrophist theory. Catastrophes could tear down as well as build up. Some geological forces were probably so destructive that they might have eliminated a good portion of the geological record. There must have been long periods of time when strata were being eroded rather than being built up. So how could you ever be certain what fossils those now vanished strata ever contained?

Who Are They?

Charles Lyell (1787–1875) was born in Kinnordy, Scotland. Although he studied law, he soon turned to science and became one of the most influential geologists of modern times. His *Principles of Geology* (published in 11 editions, beginning in 1830) had a lasting impact on Charles Darwin's evolutionary theories.

Lyell's Geological Epochs

In 1833, Charles Lyell became the first geologist to subdivide geological epochs, based on the ratio of fossils of modern mollusks to fossils of extinct mollusks in the sequences of rock strata:

➤ **Cenzoic Epoch:** 65 million to 1.6 million years ago

➤ **Paleocene Epoch:** 65 million to 55 million years ago

➤ **Eocene Epoch:** 55 million to 38 million yeas ago

➤ **Oligocene Epoch:** 38 million to 24 million years ago

➤ **Miocene Epoch:** 24 million to 5 million years ago

➤ **Pliocene Epoch:** 5 million to 1.6 million years ago

Modern geologists divide the later period of the Cenzoic Period, called the Quaternary Period, further:

➤ **Pleistocene Epoch:** 1.6 million to 10,000 years ago

➤ **Holocene Epoch, also Post-glacial:** 10,000 years ago

No to Organic Progression

Turning from the geological record to the organic one, Lyell found no more reason to sympathize with Geoffroy's views than he did with Cuvier's. Lyell was dead set against the idea of organic progression. He deplored any theory that echoed the old Aristotelian ladder of being, which proposed a "successive development of animal and vegetable life, and their progressive advancement to a more perfect state." On the contrary, he said, the development of organic life occurred in "one uninterrupted succession of physical events, governed by the laws of Nature now in operation." If you were to carefully study all the geological and fossil evidence, he said, you would have to conclude that organic change doesn't have a particular direction—lower to higher and more complex forms, for instance—but rather seems to proceed backward and forward. There were no reasonable grounds to suppose that a lower, less complex form of life couldn't succeed a higher, more complex form—a position completely antithetical to Cuvier's.

Lyell firmly rejected Cuvier's notion of four branches of organic life, all of which could be organized from lower to higher forms, with humans at the top of the heap:

> If we then examine the animal remains of the oldest formations, we find bones and skeletons of fish in the old red sandstone, and even in some transition limestone below it; in other words, we have already vertebrated animals in the most ancient strata respecting the fossils of which we can be said to possess any accurate information ….

What he's saying here is that, even in the most ancient rock formations, a vertebrate—a fish—turns up, something that wouldn't be expected if a natural progression from lower (invertebrate) to higher forms (vertebrates) actually occurred. This is especially true where humans are concerned, he said.

"The superiority of man depends not on those facilities and attributes which he shared in common with the inferior animals," Lyell said, "but on his reason, by which he is distinguished from them …." Humans wouldn't occupy their superior position without their powers of reason. It doesn't follow that the creation of man could have been the last link in a progressive chain. The transition between an "irrational to a rational animal is a phenomenon of a more distinct kind than the passage from more simple to the more perfect forms of animal organization and instinct."

Uniformitarianism: The Rules of the Game Stay the Same

Lyell decided to reconstruct a geological history of his own based on the marine fossil life he had studied and his knowledge of geology. Rather than getting trapped in the age-old question of how life originated, he concentrated on natural development. You didn't need to concern yourself with the direction of change, Lyell argued, as long as you could say something useful about how geological and organic change occurred.

Then Lyell took the next step: Why should we necessarily think that the natural processes observable on Earth today are any different than they were 10,000 or 100,000 or 1,000,000 years ago? What if you assume instead that the same physical causes have shaped the Earth and its inhabitants at all times and at the same rate? This method was an intellectual necessity—after all, if Lyell was unable to construct a history based on what he could observe in the present day, what alternative did he have? Geological formations didn't need catastrophes; you could account for all features of the Earth by the action of weathering, volcanic eruption, sedimentation, erosion, and so on. As Lyell put it, "(a)ll former changes of the organic and inorganic creation are referrible (sic) to one uninterrupted succession of physical events, governed by the laws of Nature now in operation." That is to say, the rules of the game have always remained the same: The processes at work eons ago are still at work today. This is the theory of *uniformitarianism:*

What Does It Mean?

Uniformitarianism is the theory that the natural processes that change the Earth in the present have operated in the past at the same gradual rate.

➤ All natural processes seen today have been at work throughout the Earth's history.

➤ All natural processes take place at the same gradual rate.

➤ There is a uniformity of causes and effects—the same causes will have the same effects, regardless of location or time.

➤ The effects of these processes are constant—they do not vary from one place to another or from one time to another.

Until 1859, scientists in both Europe and America had the choice of coming down on the side of Lyell and accepting uniformity of natural processes, or accepting the idea of progressive organic succession. They could not accept both. Charles Darwin changed all that by showing that the two sides were conducting the wrong debate.

Darwin's Teacher: Adam Sedgwick

A third geologist belongs in the pantheon of William Smith and Charles Lyell: Adam Sedgwick. Much of Sedgwick's fame is due to his wisdom in hiring a young field assistant one summer to help him in his geological explorations of the Welsh countryside. That field assistant's name was Charles Darwin.

Born in 1785, Sedgwick enjoyed a happy childhood, which he spent collecting rocks and fossils. In the late 1820s, Sedgwick had already won recognition for his exploration of Scottish and Welsh geology. He spent much of the following decade studying fossils in systems of rocks known as the Cambrian. Sedgwick was convinced that this system contained fossil evidence of the beginning of life on Earth because no fossils were known that were older than those of the Cambrian. His reputation soon earned him a position as a professor of geology at Cambridge University.

Like Cuvier and Agassiz before him, Sedgwick asserted that the history of Earth was one in which geological cataclysms had played a significant role. As a catastrophist, he was an opponent of Lyell, who believed that slow, gradual change—erosion, weathering, and so on—was largely responsible for shaping the Earth's surface. However, Sedgwick was open to the possibility that at least some of the "catastrophic" changes might actually turn out to be gradual. He also admitted that he was wrong in ascribing rock deposits closer to the surface, called *Upper Pleistocene* rock deposits, to the Flood. (Rock deposits closer to the surface are among those laid down last and thus are much younger than those that lie deeper.) He acknowledged instead that many of these deposits were likely to have been formed by glaciers, not floods.

What Does It Mean?

The **Pleistocene Epoch** refers to a geological period 1.6 million to 10,000 years before now. (The Pleistocene Epoch was followed by the Holocene Epoch.) During this period, much of the Earth was covered by ice. The first humans also appeared during this epoch.

Sedgwick was also perfectly willing to accept the idea that some kind of evolutionary process—then called theories of "development"—might have taken place. And he was equally amenable when it came to acknowledging that the history of the Earth must extend back extremely far in time. What he wasn't willing to give up was his conviction that a divine hand was at work in nature, which he characterized as "a power I cannot imitate or comprehend—but in which I believe, by a legitimate conclusion of sound reason drawn from the laws of harmonies of nature."

In the winter of 1831, Sedgwick had a new student in his class—Charles Darwin who had just passed his exams for a bachelor's degree from Cambridge University. Sedgwick recognized in Darwin a brilliant and curious student, so he invited him to serve as his field researcher the following summer on a geologic exploration of northern Wales. For Darwin, the trip turned into a "crash course" in field geology, one that proved a formative influence on his thinking about evolution. After leaving Cambridge, Darwin kept in constant communication with his old professor, sending him notes filled with geological descriptions that he made during his epical voyage around the world on the *H.M.S. Beagle*.

In a letter to Darwin's family, Sedgwick couldn't contain his enthusiasm for Darwin's powers of observation. He was equally impressed by the collection of South American fossils that Darwin had shipped back. "He is doing admirably in S. America & has already sent home a Collection above all praise," Sedgwick wrote. "It was the best thing in the world for him that he went out on the Voyage of Discovery ...." As we'll see, Sedgwick's excitement about his former student's prospects dimmed considerably after he read Darwin's seminal work, *The Origin of Species*. Sedgwick might have believed in the idea of evolution, but only up to a point.

The Least You Need to Know

➤ Geology came into its own as a science in the first half of the nineteenth century, thanks in large part to the work of such eminent geologists as William Smith, Charles Lyell, and Adam Sedgwick—all Englishmen.

➤ William Smith pioneered the study of strata, which, geologists discovered, occurred everywhere on Earth and contained the same sequence of organic fossils.

➤ Rejecting the idea that catastrophes such as the biblical Flood could account for significant geological change over the eons, Charles Lyell countered with a theory that the same results could just as easily have been achieved by gradual processes such as erosion and weathering.

➤ Lyell also rejected the idea of organic progression, where lower forms evolve over time into higher, more complex forms; instead, he asserted that there was no reason to think that lower forms couldn't also follow higher ones in time.

➤ To account for the organic and geologic changes that have occurred on Earth, Lyell offered the theory of uniformitarianism, which states that the same processes at work in nature today have always been at work.

➤ Adam Sedgwick exerted a great deal of influence on his star pupil, Charles Darwin, even though the two later had a falling out over Darwin's theory of evolution.

Charles Darwin and the Voyage of the *Beagle*

We've traveled a long way from speculations of the ancient Greek philosophers and the story of creation as given by Genesis. We've looked at some thinkers who believed that species never changed, and others who accepted the idea that they did change but disagreed over how that change took place. We've surveyed formalists and functionalists and catastrophists and uniformitarianists. We've seen how, little by little, evidence kept showing up—whether from explorations of the far-flung corners of the Americas or from excavations into the depths of the Earth—that forced naturalists, botanists, geologists, and zoologists to rethink some of their original conceptions of how life originated and how it had developed over long periods of time.

Now we come to the most important pioneer in the evolutionary thinking, the scientist whose name will be forever attached to the idea of evolution: Charles Robert Darwin. Darwin put forward his evolutionary theory—and very eloquently, too—in two major works: *On the Origin of Species by Means of Natural Selection* (1859) and *The Descent of Man* (1871), both of which have had a profound influence on subsequent scientific thought.

Although Darwinian theory was an unquestionably revolutionary idea, Darwin himself was not a rebel or an egocentric maverick out to take the world by storm. On the contrary, he was a meticulous and methodical observer, researcher, and scientist. It was his life's ambition to prove—or disprove—that life had descended from common ancestors and, in spreading throughout the world, had changed, turned into new species, or died out altogether.

In this chapter, we'll see how Darwin, who seemed to have little ambition as a youth besides hunting and collecting fossil and animal specimens, became interested in unraveling some of the most daunting mysteries of nature. Had he remained in his native England his whole life, however, it's unlikely that he would have collected enough facts and specimens or observed enough in the way of natural phenomena to put him on the path that he ultimately chose. He needed to have a chance to explore the world and learn how to become a scientist before he could have ever conceived of his theory. That chance came with the offer to travel for five years around the world as an assistant researcher on a British ship called the *Beagle*. It was a voyage that was destined to change Darwin—and the world.

"Shooting, Dogs, and Rat Catching"

Charles Darwin was born in Downe, Kent, in England on February 12, 1809—an auspicious day, because it was also the day that Abraham Lincoln was born. The fifth child of Robert Waring Darwin, a physician with one of the largest practices outside London, and Susanna Wedgewood Darwin, Charles was raised with all the advantages of a secure professional, upper-middle class family. The only event to mar his otherwise golden childhood was the death of his mother when he was eight.

Like so many other budding scientists whom we've considered in our story, Darwin was keenly interested in chemistry and collecting specimens. But he proved an exceptionally poor student when he was packed off to the prestigious Shrewsbury School, where the classics were emphasized. The headmaster rebuked Darwin for wasting his time with chemical experiments. Darwin was undaunted. He had no intention of giving up his two big passions in life: collecting shells, bugs, birds, minerals, maritime fossils; and hunting.

University Life

When was 16, his father sent him to study medicine at the University of Edinburgh in hopes that Darwin would follow in his footsteps. But the young man had only to witness two operations, conducted without anesthesia (a common practice in those days), to decide that he wanted nothing further to do with medicine. All the same, the two years that he spent in Scotland weren't a total loss: Darwin benefited from friendships with the zoologist Robert Grant, who introduced him to the study of marine animals, and the geologist Robert Jameson, who encouraged his growing interest in the history of the Earth.

In spite of his scientific curiosity, Darwin gave every indication that he was prepared to lead the life of a leisurely country gentleman. His father was chagrined by his son's seeming lack of ambition: "You care for nothing but shooting, dogs, and rat catching, and you will be a disgrace to yourself and to all your family." Hoping to appease his father, Darwin agreed to undertake a career in the ministry. At the start of the academic year 1828 he entered Christ's College, Cambridge, as a divinity student. But he proved no better of a student than he had been at Shrewsbury or Edinburgh, and he fell in with a group of students who treated the university as a kind of sportsman's club. Ordination as a Church of England minister held absolutely no appeal for him.

All this time, though, he had continued to be interested in the natural sciences. In the early nineteenth century, only two accepted routes were available into the scientific world: mathematics, which included the physical sciences, and medicine, which included physiology. Because Darwin hadn't mastered mathematics and the practice of medicine appalled him, a scientific career would seem to have been out of the question. However, while at Cambridge, Darwin befriended a professor of botany, the Rev. John Henslow. Henslow encouraged Darwin in his scientific studies.

Who Are They?

Alexander Freiherr von Humboldt (1769–1869) was a German naturalist and explorer who made many significant contributions to the understanding of geophysics, meteorology, and oceanography.

In his last year at Cambridge, Darwin read two books that later inspired him: Alexander von Humboldt's *Personal Narrative of Travels to the Equinoctial Regions of America During the Years 1799–1804,* and Sir John Herschel's *Introduction to the Study of Natural Philosophy*. Reading Humboldt, Darwin was gripped by the desire to see the Americas for himself; reading Herschel, he gained "a burning zeal to add even the most humble contribution to the noble structure of Natural Science."

Introduction to Adam Sedgwick

1831 was a fateful year in Darwin's life. Darwin obtained his Bachelor's degree in February. Then, with some time on his hands, he took Henslow's advice to learn geology, although he had no special interest in the subject. It was for that reason that he decided to sit in on lectures given by Prof. Adam Sedgwick. Always eager for an adventure, Darwin accepted Sedgwick's offer to accompany him on a geology tour of northern Wales in August as his research assistant. Around this time, at Henslow's urging, Darwin picked up a copy of Lyell's *Principles of Geology.* (Henslow had urged him to read it—but not necessarily to believe everything Lyell said.) Lyell's book astonished Darwin. "It then first dawned on me that I might write a book on the geology of the countries visited, and this made me thrill with delight," he would later write.

Recall that Lyell advocated a view that the face of the Earth had changed *gradually,* not catastrophically, over long periods of time through the continuing, cumulative effects of local disturbances. The same geological processes that had carved out the Earth's surface by erosion, the build-up of rock deposits, volcanic activity, and so on had been operating in the distant past and could be observed in the present. This was the theory of uniformitarianism. Lyell's work would exercise a powerful influence on Darwin's thinking when it came time to develop his own ideas about evolution.

As he traipsed through the Welsh countryside in the company of his professor, Darwin began to realize that science "consists of grouping facts, so that general laws and conclusions may be drawn from them." Although he had gotten a great deal out of his experience as an amateur geologist, he didn't see any particular reason to alter his habits. He was still anxious to get back to Kent in time for the beginning of the fall hunting season. But as soon as he got home, he found a letter waiting for him from Henslow. It was a letter that would change his life.

Henslow explained that he had been asked by Capt. Robert Fitzroy of the Royal Navy to recommend a young man to accompany him—without pay—as a naturalist on the forthcoming voyage of his ship, the *H.M.S. Beagle.* Specifically, Fitzroy was seeking "a scientific person qualified to examine the land." Henslow had thought of Darwin. Darwin's father rejected the idea, asserting that it was a "wild scheme" that certainly couldn't help Darwin's prospects as a clergyman. Only through the intercession of Darwin's uncle, Josiah Wedgewood, was the trip salvaged. The pursuit of nature, Wedgewood assured Dr. Darwin, was indeed suitable to a clergyman. Darwin, who had no intention of becoming a member of the clergy, had other ideas. In a letter to Fitzroy accepting the post, Darwin said that he expected the voyage to be a "second birth."

What Does It Mean?

Chronomatic measurement, also known as absolute dating, refers to a method of assigning a calendar year date to artifacts, fossils, and other remains. Absolute dating of artifacts from ancient cultures is one of the greatest challenges of archaeology.

The Voyage of the *Beagle*

At the age of 22, Darwin embarked on his grand adventure as an unpaid "naturalist" aboard the *H.M.S. Beagle,* a 10-gun brig, which set sail on December 27, 1831, from Davenport, England. Its mission was to study the west coast of South America and several Pacific islands, and to carry out a chain of *chronometrical measurements*—a method of dating archaeological finds—around the world. It would be five years before Darwin saw England again.

Darwin was exhilarated by his experiences aboard the brig, even though he suffered from seasickness. The exotica of the tropics captivated him. Far from being daunted by the dangers he encountered, he brazened his way through armed political rebellions, happily rode with the gauchos (South American cowboys) in Argentina, and plunged into Amazonian jungles to hunt for game and collect specimens. He even succeeded in rescuing the expedition by saving a boat from a tidal wave. In a letter to one of his sisters he exulted:

> We have in truth the world before us. Think of the Andes; the luxuriant forest of the Guayquil [sic]; the islands of the South Sea & New South Wales. How many magnificent & characteristic views, how many & curious tribes of men we shall see—what fine opportunities for geology & for studying the infinite host of living beings: Is this not a prospect to keep up the most flagging spirit?

As the voyage went on, Darwin developed increasing confidence in his own observations. He had the ability to grasp the essence of a problem and wrestle with it until he had it solved to his satisfaction. He was learning the scientific method as he went along, using the facts that he gathered to form a basis for theories that could explain them. By focusing his attention on problems involving organic and geologic change, on a continental scale, he was motivated to come up with universal laws—laws that would apply everywhere and at all times. In the five years that Darwin was away on the *Beagle,* he transformed himself into an independent and adventurous thinker ready to challenge the accepted wisdom of his age.

Strange Correspondences

By the time the *Beagle* reached Bahia, Brazil, Darwin was already busy collecting as many plants and geological specimens as possible until the ship's deck could no longer contain them all. But Darwin was not just a collector; he was a perceptive observer, capable of making connections that others in his situation probably would have missed.

One day, for instance, while walking in the Brazilian forest, he found a curious fungus that resembled one he was used to seeing in England. As he carried it in his hands, a beetle lighted on it. Darwin was amazed. The fungus that he recalled from England also held a strange attraction for beetles. Admittedly, it was a different kind of beetle, but that wasn't the point. Why, Darwin wondered, should beetles develop the same kind of intimate relationship with fungi in lands that were so geographically isolated? If English and Brazilian beetles and fungi had been separately created, why had the same link developed between them? Such correspondences were hardly unusual. Wherever he went on his explorations, he encountered similar phenomena, indicating a remarkably complex relationship connecting plants, animals, and environment.

Darwin was also struck by the capacity of organisms to adapt to the most extreme conditions. Discovering worms inhabiting pools of brine in Argentina that were seemingly incompatible with life, he wrote:

> Well may we affirm that every part of the world is habitable! Whether lakes of brine, or those subterranean ones hidden beneath volcanic mountains—warm mineral springs—the wide expanse and depths of the ocean—the upper regions of the atmosphere, and even the surface of perpetual snow—all support organic beings.

Even humans appeared far more adaptable to brutal conditions than Darwin had imagined. At Tierra del Fuego, at the very tip of the South American continent, the *Beagle* had to navigate around gigantic icebergs, many "as tall as a cathedral." When the ship put into port, Darwin was astonished at seeing the natives, who had virtually no protection against the elements. (Keep in mind when reading the following passage that Darwin was also a product of a Victorian society and blindly accepted many of its prejudices against people who were different.)

> The Fuegians slept at night on the wet ground coiled up like animals, and principally lived on shellfish or small fish they caught on hookless lines .... Whilst beholding these savages, one asks, whence have they come? What could have tempted a tribe of men to travel down the Cordillera or backbone of America ... to one of the most inhospitable countries within the limits of the globe? ... There is no reason to believe that the Fuegians decrease in number .... Nature has fitted the Fuegians to the climate and the productions of his miserable country.

Continuing with the English expedition across the Salado River of Argentina, Darwin began to remark on some striking environmental changes. He observed that the coarse high grass of the pampas changed into "fine green verdure." Initially, he thought that the change must have been due to some alteration of the soil. What he discovered was something entirely different: The new grasses had developed because of the grazing and manure from cattle—in other words, the environmental change was basically caused by humans.

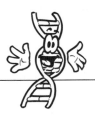

A Natural Selection

The Rio Salado is a river in northern Argentina, more than 1,100 miles long, that rises into the Andes and passes through regions devoted to mining and raising livestock.

The Human Hand in Nature

The impact of human intervention was apparent in many of the areas through which Darwin traveled. Hundreds of square miles of the plains of the Banda Oriental in Uruguay had been taken over by the prickly cardoon, a relative of the artichoke that had originally come from Europe. The promiscuous cardoon had grown so rapidly—and so densely—that it had made the area impenetra-ble to both man and beast. Here was a dramatic example of what could happen when a particular locale was invaded by a new species against which no natural enemy ex-isted. An introduction of a new species, Darwin noted, could change the balance of nature, and an established species could lose out to a new one.

Many of the rocks that Darwin examined during his travels in South America con-tained fossils of extinct species. When the expedition reached Patagonia in the south of Argentina, Darwin came across the fossil bones of an animal that resembled both a llama and a camel yet also had the build of a rhinoceros and a tapir. Darwin was cer-tain that he had found the ancestor of the llama:

> This wonderful relationship in the same continent between the dead the living, will, I do not doubt, hereafter throw more light in the appearance of organic be-ings on our earth and their disappearance from it, than any other class of facts.

In comparing the fossils of horses that he observed in South America with pictures of their North American counterparts, he was struck by how the ancient animals on both continents were more nearly alike than their living descendents. In his journal, he wrote, "I know of no other instance where one can almost mark the period and the manner of the splitting up of one great region into two well-characterized zoolog-ical provinces (North and South America)." That is to say that, even geographically separated, extinct species enjoyed a closer resemblance than their present-day descen-dents. But by what mechanism did new species replace extinct ones? At this point, rather than trying to come up with definitive answers, Darwin was mainly absorbing and observing. The ideas that ultimately coalesced to form his theory of evolution were still far in the future.

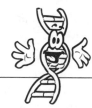

A Natural Selection

Patagonia is a region of southern Argentina, located east of the Andes Mountains, with an area of about 300,000 square miles. The region encompasses Tierra del Fuego, an archi-pelago off the southern tip of South America. Known for its sheep raising, Patagonia was first discovered by Europeans led by the Portuguese explorer Ferdinand Magellan in 1520.

Darwin made another intriguing observation about extinction. The distribution of fossils, he noted, seemed to indicate that a species first became rare before it disap-peared entirely, indicating that some change must have occurred that had weakened the species to such an extent that it could no longer survive. What had accounted for the extinction of so many species? Influenced by Lyell's theories, he did not think that geological catastrophes could have been responsible.

When he pondered phenomena like the prickly cadoon, Darwin began to realize that even the most voracious pests imaginable couldn't take over everything. But why? What prevented invading species from running amok and driving out existing species wherever they took hold? Obviously this was a crucial problem when it came to maintaining the delicate balance in nature. After all, many species laid eggs by the thousands, but most of the offspring never survived. And seeds upon the millions were carried aloft on the winds, yet the vast majority of them did not germinate. If there wasn't some mechanism for preventing these species from propagating without cease, Darwin thought, the Earth would surely long ago have been overrun. In addition, in most cases, long-established species remained fairly constant in number over time, which suggested that there were checks on even the most aggressive invaders.

As we've seen in our story so far, an understanding of organic change on Earth couldn't be achieved without also understanding how geological change came about. The two subjects were inextricably bound up in the same package—literally so, given that the fossil record is embedded in the strata of rock.

The Andes Mountains held lessons in both subjects for Darwin. For Darwin, the South American continent was a vast practical testing ground for Lyell's ideas. The Andes offered an outstanding example of how a physical barrier such as mountains or oceans could impact the development of organic life. For example, Darwin observed that the 13 species of mice he had collected on the Atlantic side of the mountain range differed completely from the five species that he had found on the Pacific side. And while crossing over the Andes in 1835, Darwin also noticed how natural forces could transform the terrain, sweeping mountains away, evicting oceans from land they'd covered for millions of years, and forcibly relocating trees and boulders and plunking them down hundreds of miles from their former homes.

Lyell must have surely been on Darwin's mind when he came on the improbable sight of petrified trees rising above a bare slope 7,000 feet above sea level. How did the trees get there? Fossil shells that Darwin found farther up, at an elevation of 12,000 feet, caused him to think that a chain of suboceanic volcanoes had disgorged enormous quantities of lava that had formed the Andes in a continual process of upheaval and fracturing. Once these same bare trees had lined the shores of the Atlantic, now 700 miles away. But successive volcanic eruptions, Darwin reasoned, had thrown the trees under water and then thrust them back up before finally hurtling them onto the mountain slopes, after which the wind and weather had gone to work on them. (Later, in Chile, Darwin experienced his first earthquake and watched the land rise before his eyes.) To one of his sisters he wrote that the Andes crossing had helped him understand "to a certain extent the description & manner of the force, which has elevated this great line of mountains."

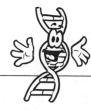

A Natural Selection

The Andes Mountain range extend, more than 5,000 miles along the western length of South America, from the Caribbean coast in the north to Tierra del Fuego in the extreme south. The Andes, with peaks as high as 20,000 feet, are second only to the Himalayas in height. These mountains also contain some of the highest permanent human settlements on Earth.

The Mystery of Mysteries

The most remarkable part of Darwin's voyage of discovery was still to come. Reaching Chatham Island, in the Galapagos archipelago off the Ecuadorian coast, he was startled by the sight of two tortoises crawling about in a desolate landscape made up of black lava, leafless shrubs, and large cacti. They were nothing like any creature he'd ever laid eyes on; it was as if they had emerged from some prehistoric era.

Lizards were particularly abundant, but Darwin wondered how they had gotten to such a remote place. It seemed improbable, but still within the realm of possibility, that in the past a few lizard eggs, encased in their shells, might have drifted from the mainland several hundred miles away. But then how did the slimy spawn of frogs find their way to these islands? There was no way that they could have survived a journey of such length. With the exception of one finch, Darwin bagged no less than 26 kinds of birds that were found nowhere else in the world. That also was true with most of the land and seashells that he collected. But that wasn't the most astonishing thing about the animal life on the archipelago.

A vice governor of the islands told him that he could distinguish from the appearance of each tortoise which island it had come from. How could this be? These islands were only 50 or 60 miles apart, after all; besides, they had the same climate and were made of the same basaltic rock. Yet, in spite of their same environment, the islands contained many different species of tortoises, lizards, and finches. The strong currents between the islands would have made it impossible for any egg or seed to migrate from one to another. The winds were never strong enough to blow birds, insects, or seeds from one to another. And very few migrant plants and animals had been able to reach the islands from the mainland. So now Darwin was confronted with two natural barriers: The Galapagos were too far away for life to have easily relocated there from the mainland, and, once there, it wouldn't be able to move to other islands.

Those few successful immigrants that somehow had settled on the islands seemed to have developed in isolation, following their separate paths until new species had been formed. But that raised another problem: If each species had been separately created—a widely held belief at the time, which Darwin shared—why should the unique giant tortoises of the Galapagos Islands resemble those of the South American mainland? Why should some upland species of ducks, in spite of the fact that they were never near water, still have webbed feet? Darwin felt that he'd come face to face with "the great fact, the mystery of mysteries—the first appearances of new beings on earth."

A Natural Selection

The Galapagos Islands (or Colon Archipelago) is a group of islands in the Pacific Ocean, about 650 miles off the coast of Ecuador (which the islands are part of). Composed of 15 large and several hundred small islands, the Galapagos are home to the unique species of tortoises, finches, lizards, and other wildlife that makes them one of the world's most important unspoiled natural preserves.

Australia and New Zealand—And Home

When the *Beagle* later reached Australia and New Zealand, Darwin witnessed another natural phenomenon that was more disturbing than anything he had seen so far. As he watched how the English settlers were behaving in their daily life, Darwin was struck by how different types of men seemed to act on one another in the same way that different species of animals did, with the stronger overwhelming the weaker. It was an upsetting and unsettling observation, suggesting that humans were not necessarily entitled to the privileged position in nature to which earlier naturalists had assigned them.

After five years, the *Beagle* returned to England, landing in Falmouth on October 2, 1836. Darwin now turned to the task of reporting on his findings, summarized in an enormous five-volume project titled *The Zoology of the Voyage of H.M.S. Beagle, Under the Command of Captain Fitzroy, R.N., During the Years 1832 to 1836.* In it, Darwin covered all the life forms, existing and extinct, that had captured his interest throughout his journey—fossils, mammals, birds, fish, and reptiles. But the books offered readers only a comprehensive collection of observations and commentary, written by Darwin with the help of several experts in various fields. What all these facts added up to, no one could say at this point—not even Darwin himself. The revelatory message of the theory of evolution wouldn't be heard for several years to come.

The Least You Need to Know

➤ Charles Darwin, whose theory of evolution changed the world, was born in 1809; although he demonstrated enthusiasm for nature and loved to collect fossils and shells, he was an indifferent student.

➤ At Cambridge University, Darwin's friendship with professors of botany and geology led to his commission as a research assistant attached to a scientific voyage of the British ship the *H.M.S. Beagle.*

➤ The five-year journey on the *Beagle* around the world gave Darwin exposure to the world and the experience to allow him to pursue his lifelong goal of discovering how life originated and changed over time.

➤ Observations in his travels fed Darwin's growing conviction that newer species, because they are more aggressive, are capable of supplanting established species in a particular locale.

➤ Darwin noted, however, that no species, even the most aggressive, can overrun the Earth unchecked. Some mechanism, he reasoned, comes into play to stop invading species from completely taking over from established species.

➤ According to Darwin, the creatures inhabiting the Galapagos must have pursued their own avenues of development, even though they once must have shared a common ancestor.

Chapter 9

Nature's Winners and Losers

In This Chapter

➤ The development of Darwin's thinking about species

➤ The influence of Malthus's theory of population on Darwin

➤ The survival of the fittest

➤ How and why natural selection works

Once Darwin got back to England, he set about compiling the notes that he had made during his five-year journey around the world on the *Beagle*. He had seen a great deal—volcanic eruptions, extinct fossils, regions swallowed up by noxious plants, petrified trees hundreds of miles from the ocean where they once grew, and unusual species of tortoises, finches, and lizards—but how to make sense of it all? "The mystery of mysteries" still resisted Darwin's solution. What conceivable theory could Darwin come up with that would encompass so many different phenomena?

For the first few years after his return, Darwin concentrated on determining how environmental influences act on species. He took it as a given that animals and plants are constantly competing for scarce resources and that some individuals are winners in the contest while others are losers. He also assumed that, over time, congenial environments can abruptly change into inhospitable ones: Seas recede, jungles are transformed into desert, and invading species overrun native species. Again, some individuals are winners because they manage to adapt to the new conditions, and others, unable to cope, fall by the wayside.

Darwin wasn't attempting only to derive a theory that would explain how species met these environmental challenges and adapted (or not) as a result. It also sought to explain how the same adaptive process might eventually lead to the creation of a new species. That Darwin succeeded in conceiving such a theory is remarkable enough, but what makes that theory so extraordinary is that it provided a plausible description of how evolution actually works in nature.

Evolution of a Theory

When Darwin returned home after 5 years abroad, he was 27 and still had another 45 years to live. He settled in London, where he finally made the acquaintance of the geologist Charles Lyell, whose work he had admired since his days at Cambridge. Their friendship was to have a lasting influence on the course of Darwin's future work. And work was really all that Darwin wanted to do. The claims of marriage, children, and professional obligations all seemed to him to be many distractions. But as a lifelong hypochondriac, he was able to claim ill health as an excuse to avoid almost all the professional—and domestic—demands made on him, allowing him time to pursue his scientific investigation unimpeded.

Once Darwin had put his collections in order and prepared his journal of the voyage for publication, he resumed work on the problem of the origin of the species, hoping "to throw some light on the whole subject." To his close friend, the botanist Sir Joseph Dalton Hooker, he wrote, "I have been ever since my return engaged in a very presumptuous work and I know no one individual who would not say a very foolish one ...." The task that he'd set for himself took another 20 years. From 1836 on, he continued to work on the problems he had first grappled with on the *Beagle,* collecting facts, experimenting, reading, and abstracting.

His experiences abroad had convinced him of the fact of organic evolution. The facts he had gathered, he said, "indicated the common descent of the species." But how did this evolution by descent come about? He was still trying to figure it out, but for the time being he kept collecting facts "on a wholesale scale ... facts which bore in any way on the variation of animals and plants under domestication and nature." Darwin needed all the facts he could lay his hands on, too, if he was to prove his case convincingly to a skeptical public. Of course, that implied that he had a case to make. And he still didn't.

Tying the Present to the Past

One of the problems Darwin struggled with was tying the present and the past together by establishing the relationship of existing species with extinct ones. He wrote in one of his first abstracts:

> When on board *H.M.S. Beagle,* as naturalist, I was much struck with certain facts in the distribution of the inhabitants of South America, and in the geological relations of the present to the past inhabitants of that continent. These facts

seemed to me to throw some light on the origin of species—that mystery of mysteries, as it has been called by one of our greatest philosophers .... After five years' work I allowed myself to speculate on the subject, and drew up some short notes ...."

Before he advanced his own theory, Darwin summarized the various theories that were still in vogue in the middle of the nineteenth century.

Until recently the great majority of naturalists believed that species were immutable productions, and had been separately created. This view has been ably maintained by many authors. Some few naturalists, on the other hand, have believed that species undergo modification, and that the existing forms of life are the descendants by true generation of pre-existing forms.

There were two possibilities: Species were either fixed, or they changed and evolved from earlier forms. Darwin had no doubt about where he stood. Species did undergo modification, he said, and individual members of a species varied from one to the other. All the same, Darwin didn't underestimate the difficulties that he was up against: "Our ignorance of the laws of variation is profound. Not in one case out of a hundred can we pretend to assign any reason why this or that part has varied."

Fundamental to Darwin's thinking was the conviction that the same processes at work in the past are still at work today. The present was the key to the past. Remember that this was Lyell's position, too. And, like Lyell, Darwin didn't think it necessary to fall back on the idea that frequent catastrophes had given the Earth's surface its present shape. Even a stream, if it flows long enough, will tend to wear away its banks and, over time, can carve out a new valley. If landscapes underwent change so slowly, Darwin thought, why shouldn't the same gradual process apply to organisms?

But many questions remained unanswered. What mechanism in the environment, for example, could transform a single original population into a dozen different species? That was apparently what had occurred in the Galapagos. Why did each island have its own distinct flora and fauna? Tortoises differed from one island to another. Finches on one island had different forms, habits, and diets from finches on other islands. Some finches, for instance, had sharp and narrow beaks, while others had curved and stubby beaks. They were clearly distinct species. Plants exhibited the same kind of differences, depending on which island they grew.

These differences seemed to suggest to Darwin that all existing species must have at one point had common ancestors. But how could species so close in time and space turn out to be so different? Alternatively, how could species so widely separated geographically and geologically (think of the upland ducks or the tortoises on the mainland of South America and those on the Galapagos) be so similar? To explain both the differences and the similarities between species otherwise so far removed, Darwin realized that he would have to come up with an entirely new theory.

In his efforts to make sense of the many observations he had made during his expedition, Darwin recalled the way in which species—among them foreign invaders such as the prickly cadoons and English settlers of Australia and New Zealand—struggled with one another for supremacy. Several naturalists already had noted the competition between the species. You may remember that Linnaeus had referred to nature as a "butcher's block," while a Swiss botanist named DeCandolle, whose writings Darwin had read, declared, "All the plants of a given country are at war with one another. The first which establish themselves by chance in a particular spot tend, by the mere occupancy of space, to exclude other species …."

Lyell, too, had remarked on the phenomenon. There was, he said, a "universal struggle for existence" in which "the right of the strongest eventually prevails, and the strength and durability of the race (read species) depends mainly on its proliferations." By "proliferations," Lyell was referring to the capacity of a species to reproduce and propagate a great number of offspring, ensuring that it would keep what it had won on the battlefield with its competitors.

The Creative Uses of Competition

Yet, what Darwin realized was that Lyell had gotten only one part of the picture by focusing on the destructive aspect of struggle as an instrument by which weaker species were eliminated by the stronger. In Lyell's conception, the only species that were involved in the conflict were existing species. What Darwin realized was that this same process doesn't have to be exclusively a *destructive* one. It can also be a *creative* one: Competition can lead to the development of new species. This was an idea that had never occurred to the geologist.

Assuming that competition could lead to both destructive and creative results, Darwin now had to figure out the solution to two fundamental problems. First, he had to explain how various forms of animals and plants appeared. This would mean tackling the origin of life on Earth, a challenge that many scientists had tried to dodge. Second, he had to explain why certain species were able to establish themselves at the expense of others.

Darwin's Epiphany

Then, by chance, Darwin stumbled onto the solution in October 1838. It happened in one of those serendipitous bursts of illumination. This is how Darwin recorded the incident:

> In October 1838, that is, fifteen months after I had begun my systematic inquiry, I happened to read for amusement Malthus on *Population,* and being well prepared to appreciate the struggle for existence which everywhere goes on from long-continued observation of the habits of animals and plants, it at once struck me that under these circumstances favourable variations would tend to be

preserved, and unfavourable ones to be destroyed. The results of this would be the formation of a new species. Here, then, I had at last got a theory by which to work.

"A theory by which to work"—it was a crucial moment.

Who Are They?

Thomas Malthus (1766–1834) was professor of political economy and modern history at the college of the East India Company at Haileybury from 1805 until his death. His principal work was *An Essay on the Principle of Population* (1798). He was also the author of *An Inquiry into the Nature and Progress of Rent* (1815) and *Principles of Political Economy* (1820).

An Economic Sermon No One Wanted to Hear

Who was Malthus? Thomas Malthus was a British economics professor who is best known today for *An Essay on the Principle of Population,* published in 1798. His *Essay* read more like an economic sermon about human behavior and its imperfectability than a social analysis, although it was that as well. Malthus proposed the idea that population tends to increase faster than the supply of food available for its needs. If more food is produced to meet the needs of the population, people will be tempted to have more children, and the unintended result will be a further rise in population. That, of course, only puts you back where you started, Malthus said, without enough food to sustain the expanded population.

What happens if the growth of population outstrips the food production by too much? Well, said Malthus, disaster strikes. If too many people are competing for too few resources, nature will intervene to rectify the situation in the form of famine, disease, and war. Ultimately, Malthus asserted, all schemes that you could devise to avert disaster are doomed (although, as you'll see, that didn't stop him from proposing some solutions himself), mainly because, as time goes on, the human population continually tends to *increase geometrically,* while man's capacity to increase resources *increases only arithmetically.*

What Does It Mean?

Add 4 and 4, and you get 8—that is **arithmetic increase.** Multiply 4 by 4, and you get 16—that is **geometric increase.** Add 4 to 8, and you get 12; multiply 16 by 4, and you get 64. You can see that it doesn't take long to reach staggeringly high numbers by means of geometric increase.

In a society that equated progress with high fertility, Malthus's grim depiction of the future was unwelcome news. (It's not for nothing that people who became acquainted with Malthus's theory called economics the "dismal science.") But Malthus was basing his theory on evidence that living conditions in nineteenth-century England were in decline. He attributed this decline to three factors:

➤ Too many young people being born

➤ The inability to produce sufficient resources to keep up with the increasing human population

➤ The irresponsibility of the lower classes

In spite of the class biases that colored Malthus's ideas, his work had some far-reaching consequences. For one thing, he introduced a new way of studying social phenomena. You could agree or disagree with his theory, but the only way of proving it right (or wrong) was to actually conduct a rigorous demographic study to find out whether the world's supply of people was increasing faster than sufficient resources were being produced to sustain them. In this sense, Malthus can be considered the pioneer in what is known as demographic studies, which involves examining a given population to track behavior, trends, or preferences. (Political polling and advertising are just two uses to which demographic studies can be put.)

Survival of the Fittest

Now let's return to Darwin. After he had read Malthus's *Essay,* Darwin was immediately struck by how pertinent the English economist's ideas were to some of the observations that he had been musing over ever since his return to England. It seemed to him that, in describing a situation in which too many people must compete for too few resources to survive, Malthus was talking about the very same phenomenon that DeCandolle and Lyell had remarked upon. Darwin later wrote:

There is no exception to the rule that every organic being naturally increases at so high a rate, that if not destroyed, the earth would soon be covered by the progeny of a single pair. Even slow-breeding man has doubled in twenty-five years, and at this rate, in a few thousand years, there would literally not be standing room for his progeny.

Malthus's theory, Darwin decided, could apply to any population, not only humans—monkeys, mice, tapirs, and even prickly candoons. Unlike Malthus, though, Darwin wasn't trying to find some kind of deeper social or moral significance in this phenomenon; he was just describing things the way they appeared to work in nature.

As Darwin had seen for himself in South America, plants and animals produce far more offspring than can survive. Species live in a competitive environment. Inevitably, different species—and, more often, individual members of the same species—have to struggle for scarce resources among themselves. Some individuals have a greater chance to survive (and, therefore, have a better chance of reproducing and having offspring) than others. The question that Darwin now had to confront was how the process actually worked. What endowed some individuals with the capacity to compete successfully? The answer to that puzzle turned out to provide a bonus: the answer to how new species could emerge from old ones.

Variation: The Key to the Puzzle

The idea that variation was somehow the key to the puzzle had been around for a long time. Lamarck, for instance, had focused much of his work on determining how favorable traits (longer necks in giraffes and thicker coats on dogs in winter climates) could enhance chances for survival. But, as we've said, Lamarck was mistaken in confusing superficial traits (involving appearance, for example) with the genetic makeup of the individual.

You can build up your muscles, if you choose to exercise a lot—that's appearance. You may have athletic or musical talent, or you may be very smart—that's genetic. Whether you develop your intrinsic talents is up to you. Just because you have a genetic ability doesn't mean that you'll be successful at it. If you have a susceptibility to an illness, such as diabetes or depression, that's genetic, too. You can treat the illness, but you can't determine whether you'll have the illness in the first place. Lamarck failed to distinguish between a genetic inheritance that can be developed or refined (appearance or talent) and a genetic inheritance that governs who you are (body shape, intelligence, susceptibility to disease, and so on).

Again, we have to emphasize that in the middle of the nineteenth century, the mechanism for genetic inheritance wasn't known. What was known, however, was that certain distinctive traits—albinism, for instance, or an extra finger or toe—could recur disproportionately over successive generations in a single family. Lamarck had been mistaken to insist that beneficial traits that developed over the lifetime of an individual—that is, traits that could enhance his chances for survival—could be passed on to his offspring. But even if Lamarck had gotten it backwards, he had hit on something important.

Species did respond to environmental changes, as Darwin had witnessed many times in his travels through South America and among the Pacific atolls. If the changes were drastic enough, they forced organisms to either adapt or die. Those traits (variations) that helped the organisms to adapt in the new environment were the ones that made them the winners in the competitive struggle. Their brothers, sisters, and cousins who didn't have these same beneficial variations eventually weakened and were overwhelmed by the species with the superior traits. This is what is meant by *survival of the fittest.* (Remember Darwin's observations that fossils of extinct species tend first to become rare and then to disappear altogether. Darwin had now found a hypothesis that might account for such an effect.)

What Does It Mean?

Survival of the fittest refers to the idea that, because of favorable traits, some individuals are better able to survive in particular environments than individuals without those traits.

But how, you might well ask, did these variations— favorable and unfavorable alike—occur in the first place? After all, members of a species suddenly confronted with a dramatic alteration in its environment— a flood, a sudden drought, or a prolonged drop in temperature—couldn't have anticipated what would happen. The variations that would help particular members of the species adapt successfully to the environmental change were already present. Perhaps they wouldn't have expressed themselves if the change had never happened.

The Advantage of Mistakes

Well, we've already suggested an answer to why some organisms have positive traits for survival and others do not. In a word, Darwin said, variations occur because of *mistakes.* Darwin conceived of life as a series of successful mistakes.

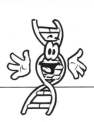

A Natural Selection

Man couldn't have attained his present "dominant position" in the world without his hands, said Darwin. But hands could not have become useful for making tools or hurling rocks unless man first learned how to become a biped, walking on two legs instead of on all four limbs.

Descent always involves modification. Parents don't produce exact copies of themselves. (That would be cloning.) Each time an offspring is produced, errors creep into the process. (We now know that these errors—those that are beneficial, those that are destructive, and those that don't mean much one way or another—are caused by genetic *mutations*.) The slightest advantage to any one offspring, at any age or during any season, will give it an edge over those it competes with. That means that those organisms with the advantage will likely come out the winners. As winners, they will be the ones more likely to reproduce and bear more offspring, passing on in the process the traits that gave them the advantage so that they can survive in a particular environment.

If one variation enhances the ability of an individual to reproduce, then that variation will multiply better than other variations. Over a long period of time, the variation will become dominant. In the end, a new form—a new species—will emerge. This is what is known as natural selection. As Steve Jones puts it in his book, *Darwin's Ghost:* "Information cannot be transmitted without loss … To reproduce in succession an original again and again is to make—to evolve—something new. What went in emerges transformed by errors of descent, the raw material of biological change."

The idea of natural selection—of errors in the form of slight, successive, and ultimately beneficial advantages creeping in over long periods of time—fits well with Darwin's belief about how nature works generally. You don't need dramatic leaps—worldwide floods and such—to explain geology, when gradual processes (the slow grinding away of a stream or an unrelenting desert wind) will do just as nicely. Similarly, you don't need drastic leaps to explain how evolutionary change within species can occur. Once you've accepted that premise, you are well on your way to understanding how this gradual, cumulative process can explain how new species are created as well.

What Does It Mean?

Differences in the offspring of an organism are called **mutations.** Although we generally think of mutations as harmful, such as birth defects, a mutation is simply an alteration of the genetic information of an organism. Some mutations are positive: A child might be stronger, more agile, or more intelligent than his parents.

This is how Darwin describes natural selection in *The Origin of Species:*

> [I]t may be asked, how is it that varieties, which I have called incipient species, become ultimately converted into good and distinct species, which in most cases obviously differ from each other far more than do the varieties of the same species? … All these results … follow inevitably from the struggle for life. Owing to this struggle for life, any variation, however slight and from whatever cause proceeding, if it be in any degree profitable to an individual of any species, in its infinitely complex relations to other organic beings and to external nature, will tend to the preservation of that individual, and will generally be inherited by its offspring. The offspring, also, will thus have a better chance of surviving, for, of the many individuals of any species which are periodically born, but a small number can survive. I have called this principle, by which each slight variation, if useful, is preserved, by the term of Natural Selection ….

A Natural Selection

The term *evolution* is not synonymous with *natural selection*. Evolution refers to changes over time in the genetic structure of a population. Natural selection specifies one way in which genetic change may occur.

Think of natural selection as a window that not everyone can get through. If you are the lucky one with all the favorable attributes to get through that window, then you will probably be the one to reproduce. But then suppose that your environment changes, and the window shifts to another position. (If it shifts too far, though, the whole population will become extinct.) Now the attributes that helped in getting you, or your competitor, through the window in the old position no longer are of any benefit. A different set of attributes is needed to climb through the window in its new position, and those who have them will be the ones to reproduce.

Obviously, if we can apply the same principle to all other forms of life, an exception can't be made for humans. This is how Darwin put it:

> In looking at Nature, it is most necessary ... never to forget that every single organic being around us may be said to be striving to the utmost to increase in numbers; that each lives by a struggle at some period of its life; that heavy destruction inevitably falls either on the young or old, during each generation or at recurrent intervals.

If the destructive forces (famine, war, pestilence, and so on) designed to keep a population from expanding diminish just a little, Darwin said, then "the number of the species will almost instantaneously increase to any amount." Then Darwin goes on to say this:

> Can it, then, be thought improbable, seeing that variations useful to man have undoubtedly occurred, that other variations useful in some way to each being in the great and complex battle of life, should sometimes occur in the course of thousands of generations? If such do occur, can we doubt ... that individuals having any advantage, however slight, over others, would have the best chance of surviving and of procreating their kind?

Here are the basic principles of natural selection:

➤ Species of all kinds are engaged in a competition for resources.

➤ Some organisms will be better able to compete than others.

➤ The organisms that possess variations that give them an advantage over their competitors—members of the same species or a different species—are more likely to survive.

➤ The organisms that live in a particular environment where that variation enhances their chances for surviving and multiplying at the expense of their competitors will be the "best-adapted" forms.

➤ The best-adapted forms (as long as there is no cross-breeding) are those that are more likely to reproduce and have offspring with the same variation.

➤ Given sufficient time, populations originating from the same ancestral stock but forced to respond to different environmental conditions and competitors might develop new characteristics, resulting in the creation of a new species.

But what about those intermediate species that had given so much trouble to so many scientists over the centuries? Were there actually species that represented a transition between the old and new species? And if there were, what happened to them? Darwin made short work of the problem:

> We have seen that species at any one period are not indefinitely variable, and are not linked together by a multitude of intermediate gradations, partly because the process of natural selection will always be very slow, and will act, at any one time, only on a very few forms; and partly because the very process of natural selection almost implies the continual supplanting and extinction of preceding and intermediate gradations.

We don't see evidence of intermediate species, Darwin is saying, because they've become extinct, vanquished by more successful usurpers.

The Double Action of Selection

The significance of Darwin's idea of natural selection lies in "the double action" of the selective process. Two sets of environmental factors come into play in the struggle for survival for both animal and plant populations—physical and biological. The physical environment includes climates, soil, water, and so on, while the biological environment includes food supply, predators, and competitors. These two types of environmental influences complement each other, as Darwin observed on the Galapagos.

On the mainland of South America, Darwin saw, organic life hadn't changed dramatically. When change did occur, it seemed to take place very slowly and over a fairly large geographical area. That was in total contrast to the islands of the Galapagos, where species seemed to have undergone very pronounced changes that had occurred very close to one another. Tortoises, for instance, had developed into different species on islands that were within sight of one another.

Now, in natural selection, Darwin had a theory to account for why change had come about so differently on the mainland and on islands 700 miles away. On the mainland, Darwin said, natural selection was coarser and much less precise: The climate hadn't varied as much, and far more space was available, as were a multitude of different environments (mountains, desert, and jungle) in which species could thrive. In the Galapagos, however, the balance of nature was much more precarious: Each island called forth advantageous variations that were custom-tailored to a very specific environment. One island, for example, favored seed-eating birds, while another proved far more hospitable to insect-eating birds.

In essence, Darwin said, organisms evolved to fit their environment based on natural selection. If a mutation was good, meaning that it helped the organism fit comfortably into its environment, the organism survived long enough to contribute the favorable variation (or mutation) to the gene pool of its offspring. If the mutation was bad, on the other hand, the organism was more likely to die before it had the chance to contribute to the gene pool. These mutations slowly produced the world that we know today over billions of years. Lizards, for instance, are "fit" to survive in their environment mainly because they are green and can be easily camouflaged by the surrounding vegetation. If they were any other color, they could be easily spotted and eaten by predators. A purple creature set down in a jungle or forest wouldn't last long for the same reason.

A Natural Selection

A species can originate but once; biologists are lucky to see it happen.

Faltering at the Finish Line

By the end of 1839, Darwin had already gained most of the fundamental insights into evolution that ultimately found their way into his seminal work, *The Origin of Species*. However, he felt burdened by outside commitments. A year earlier, he had accepted the position of Honorary Secretary to the Geographical Society. He began to complain about how the demands on his time were interfering with his work on "the species question." In a letter to Charles Lyell, he wrote:

> I have lately been sadly tempted to be idle … as far as pure geology is concerned—by the delightful number of new views which have been coming in thickly and steadily—on the classification and affinities and instincts of animals—bearing on the question of species. Note-book after note-book has been filled with facts which begin to group themselves clearly under sub-laws.

He spent another five years still working out many of the details before he had even the basic outline for *Origin*. And yet another 15 years passed before the book itself appeared. There's even some reason to think that he might have delayed further in getting his theory of evolution out to the world if events hadn't conspired to cause him to rush it into print. But that's a story we'll save for our next chapter.

The Least You Need to Know

➤ More organisms are born than can possibly survive because of the demand that they place on available resources.

➤ Offspring are similar but not identical to their parents.

➤ Variations that enhance an organism's capacity to survive and multiply will take over and supplant less favorable variations in a population.

➤ Over long periods of time—often millions of years—the cumulative effect of these favorable variations in a given population may lead to the creation of a new species.

Chapter 10

The Debut of Evolution

> ## In This Chapter
>
> ➤ Alfred Russell Wallace's efforts to understand evolution
>
> ➤ Wallace's contribution to evolutionary theory
>
> ➤ Darwin's *The Origin of Species*
>
> ➤ Reaction to *Origin*
>
> ➤ Religious and scientific objection to Darwinian theory

Darwin was convinced that he could find an explanation that could encompass the variation of species in the context of gradual geological change. At the same time, he hoped to be able to account for how new species could emerge. The key to new species, Darwin believed, was variation. But how did variation lead to the development of species never before seen on Earth? He was determined to put off publication of a book propounding his views until he was certain that he had cleared up as many problems as possible. However, events conspired to force Darwin's hand to publish his theory much faster than he had intended and in a more abbreviated form. The book was one of the most important ever written: *The Origin of Species*.

By an amazing coincidence, another naturalist (and fellow Englishman), Alfred Russell Wallace, had developed a theory of evolution that bore an uncanny similarity to Darwin's. Although Darwin would be forever afterward associated with the theory of evolution, Wallace must be considered an equal contributor to it.

Once the theory saw the light of day, regardless of authorship, it aroused intense passions—pro and con. Religious objections to the theory focused on its emphasis on the blind action of natural selection in organic development. God seemed to have no place in Darwin's universe. Scientists, too, found fault in the theory, pointing out, often with some justification, that it was full of gaps and inconsistencies. If variations persisted in organisms only because they were beneficial for survival, as the theory proposed, how could you explain the persistence of secondary sexual characteristics (the colors of male peacock feathers, for instance), even though they didn't seem to help peacocks survive?

Although Darwin had foreseen that his book would create controversy, he tried to remain above the conflict and continue his work in peace. At his death, many of the problems raised by critics had not yet been satisfactorily resolved. That task fell to the scientists who followed Darwin.

The Bombshell

In June 1858, while he was still developing his theory of evolution, Darwin received a letter from Alfred Russell Wallace, an English naturalist then working in Malaya, titled "On the Tendency of Varieties to Depart Indefinitely from the Original Type." Darwin and Wallace had corresponded over the last couple of years because both were interested in "the species question," but they had never met.

In 1855, Charles Lyell had shown Darwin an essay that Wallace had sent to him, entitled "On the Law Which Has Regulated the Introduction of New Species." The essay constituted a statement of Wallace's belief in evolution. Darwin, however, gave the essay little attention. Perhaps if he had, he might have been more prepared for the shock that Wallace's letter caused him. When Wallace wrote to Darwin, he had no inkling that Darwin was working on a theory based on natural selection, or vice versa. That was what made Wallace's letter all the more extraordinary. Wallace had apparently reached the same conclusions about natural selection that Darwin had, completely independently. "I never saw a more striking coincidence," Darwin marveled, "if Wallace had my MS sketch written out in 1842 he could not have made a better short abstract .... Even his terms now stand as heads of my chapters."

Evolution Acquires a Second Father

Alfred Russell Wallace was a true Renaissance man—naturalist, evolutionist, geographer, anthropologist, social critic, and theorist. He was born in 1823 into a middle-class family of modest means. To earn a living, he was forced to go into the family business, but he still found time to pursue the natural sciences, his true love. Inspired by a newly published book called *A Voyage Up the River Amazon*, Wallace decided to leave the family business—and England—behind, and set out with a friend on an expedition to South America. His plan was to see another part of the world and to collect specimens. His real objective, though, was to investigate the causes of organic evolution.

Wallace was particularly influenced by Charles Lyell's *Principles of Geology*—the same book that had enthralled Darwin—as well as a popular book called *Vestiges of the Natural History of Creation,* by Robert Chambers. Lyell's work, which had proposed that the natural processes observable today are the same that had always been in force, instilled in Wallace an appreciation of how the cumulative effect of gradual, continuous processes could result in long-term change. *Vestiges,* an early effort to explain biological evolution, was riddled with errors. One critic called it "an uncomfortable compound of scientific observation and ill-supported hypothesis." It still managed to sell like hotcakes. Chambers, who argued persuasively against both creationism and Lamarckism, must have had some idea of the stir that his work would cause because he had first kept his name off the cover, publishing the book anonymously.

For all its flaws, *Vestiges* excited Wallace's fertile imagination. Perhaps, he thought, he could demonstrate that evolution did take place, but in a more logical and credible way. Perhaps, he thought, he could trace back different species in different parts of the world all the way back to their ancestors in an attempt to show that, no matter how little they might resemble one another now, they all shared a common line of descent.

Wallace's Hunt for the Mechanism of Evolution

Wallace decided to study *phyla* (groups of related organisms) to see whether he could find a similarity of body structures. This was the same idea that Etienne Geoffroy St. Hilaire had promoted in the previous century; Geoffroy called them homologous structures. That whale flippers, bat's wings, and human hands all turned out to have the same underlying structure said something significant about how these species, though quite different today, might once have come from the same ancestor.

To carry out his study, Wallace focused on two areas—geography and environment. After all, geography could act either to help a species expand or to put the brakes on it. You may recall that Darwin discovered that the Andes mountain range acted as a barrier, cutting off the ability of many species to spread farther. The species of mice on the Pacific side of the Andes were not found on the Atlantic side. By the same token, the 700 miles of water separating the Galapagos from the South American mainland also posed an insurmountable barrier for species to expand—in either direction.

Second, Wallace wanted to examine how the environment in which species lived influenced their ability to adapt. Needless to say, this was hardly a

What Does It Mean?

A **phylum** is a major category, or taxon, of organisms with a common design or organization, even though structural details may differ because of evolution. It is an arbitrary category, based on observation, theorizing, and guesswork. It is assumed that all organisms in a single phylum came from a common ancestry.

simple task. To get it done, he needed to familiarize himself with everything he could about each region he was studying: its ornithology (bird life), entomology (insect life), primatology (primate life), ichthyology (fish life), botany, and physical geography. For all his effort, though, he was unable to figure out the actual mechanism of evolutionary change.

Wallace's Feverish Epiphany

By early 1852, in ill health, Wallace felt that he had no choice but to leave South America and return to England. He didn't stay there for long. With no immediate opportunities for employment, Wallace decided to resume his research, this time in what was then known as the Malay Archipelago. (Today, it's the Indonesian Archipelago.) He arrived in Singapore in late April 1854, to begin what would turn out to be the defining period of his life. During the nearly eight years that Wallace spent in Indonesia, he undertook an astonishing number of expeditions—almost 70 in all, for a combined total of some 14,000 miles of travel. He visited every important island in the archipelago at least once, in the process acquiring a collection of 125,660 specimens, including more than a thousand species never seen before by scientists.

A Natural Selection

Alfred Russell Wallace's account of his Indonesian experiences, *The Malay Archipelago*, ranks as one of the most celebrated scientific travel books. It features his adventures capturing birds of paradise and orangutans, his encounters with native peoples, and recollections of what it was like to be among the first Europeans to live in New Guinea.

Throughout his explorations, Wallace kept returning to the question that had obsessed him in the Amazon: How did species evolve? In February 1858, Wallace found himself on one of the Moluccas, suffering from a bout of malaria. As he nursed his illness, he came across Thomas Malthus's essay on population—the same essay that had crystallized in Darwin's mind how evolution might work! Wallace was just as struck by Malthus's theory as Darwin had been. Here, he thought, was a mechanism that might also be responsible for long-term organic change. Wallace had independently arrived at the ideas of the survival of the fittest: Those individual organisms that are best adapted to their environment—that is, the fittest ones—have a better chance of surviving and passing along their traits to progeny.

As soon as he recovered, Wallace wrote an essay elaborating on his discovery. He called it "On the Tendency of Varieties to Depart Indefinitely from the Original Type." This is how Wallace described the mental leap that led to his discovery:

> In all works on Natural History, we constantly find details of the marvelous adaptation of animals to their food, their habits, and the localities in which they are found. But naturalists are now beginning to look beyond this, and to see that there must be some other principle regulating the infinitely varied forms of animal life. It must strike every one, that the numbers of birds and insects of different groups, having scarcely any resemblance to each other, which yet feed on the same food and inhabit the same localities, cannot have been so differently constructed and adorned for that purpose alone.

So Wallace had followed the same path that Darwin had taken. When the two recognized that competition was constant in nature (the idea of the survival of the fittest), they then had to discover the mechanism that would allow some organisms to win the competition while eliminating others. The mechanism was natural selection.

When Wallace was finished with his essay, the first person he could think of to send it was Darwin. Who else was as interested in the "species question"? Wallace expressed his hope that Darwin would read it and then bring it to the attention of Lyell for his comments. How was Wallace to suspect that Darwin had been pursuing research along the same lines as he had for the last 20 years and that Darwin, too, had been as inspired by Malthus?

The Origin of the Origin

As soon as he read Wallace's letter, Darwin was understandably upset and perplexed. He got in touch with Lyell to ask his advice on how he should handle what was obviously a very awkward situation. Generously, Darwin had decided that he couldn't ignore Wallace's work, however much it might threaten to preempt his own. Instead, he generously offered to send Wallace's manuscript to a reputable scientific journal. "I would far rather burn my whole book than that he (Wallace) or any other man should think I had behaved in a paltry spirit," he said.

His good friends, Lyell and Joseph Hooker, suggested a compromise. Why shouldn't the world learn about what both men had done? The best forum, they agreed, would be at the next meeting of the Linnaean Society, a prestigious scientific organization. Wallace's essay and some

A Natural Selection

The Moluccas, in the easternmost part of the Indonesian archipelago, consist of about 1,000 islands, spread across an area of about 328,000 square miles, 90 percent of which is ocean. It is Indonesia's most remote province.

unpublished writings of Darwin's on evolution could be presented at the same time. (The Linnaean Society took its name from Carl Linnaeus, the great Swedish taxonomist.) No one, however, troubled to obtain Wallace's permission, and he learned about the presentation only after it was over.

On July 1, 1858, the Linnean Society met to hear the organization's secretary read the papers of both men. Darwin wasn't present, and Wallace was still in Indonesia, oblivious of what was being done in his name. The significance of the papers was not lost on many of the scientists in the audience. They understood that they were living through a historic moment, even if they couldn't quite grasp just how revolutionary the papers actually were. In fact, they had come expecting to hear a paper by a scientist named George Bentham, reaffirming the venerable idea that the species were fixed; instead they ended up hearing the exact opposite! Knowing that he was beaten even before he could mount the podium, Bentham at once withdrew his own paper.

Darwin Readies to Take Evolution Public

Now that it was apparent that he no longer enjoyed a monopoly on natural selection, Darwin realized that he couldn't delay any longer in getting his theory into print. With Lyell and Hooker's encouragement, he decided to rewrite his journal, making it suitable for publication. Ever since his return to England, he had been struggling to finish a much larger and more comprehensive book about evolution (which, in fact, was never completed). Instead, because of the unexpected change in circumstances, he was compelled to produce a more compact, readable, and, ultimately, probably more successful work—*The Origin of Species*.

Even so, he was reluctant to put his theory into print; there were several problems that he had yet to fully address. In addition, the theory needed further refinement. Published prematurely, he feared, the book might do far more harm than good. If his book were to ever get published, though, Darwin realized that he would have to admit to the problems that he hadn't been able to solve and try to work around the gaps in his knowledge. To his credit, Darwin made no attempt to prove things that he could not prove; instead, he simply presented the plausible assumptions on which he based his conclusions for his readers to judge.

> To treat this subject at all properly, a long catalogue of dry facts should be given; but these I shall reserve for my future work .... It is hopeless to attempt to convince anyone of the truth of the proposition without giving the long array of facts which I have collected, and which cannot possibly be here introduced ....

Even with all the facts at his disposal, he was aware that he wasn't going to be able to satisfy his critics. Still, with the urging of his friends, he went to work.

Uncharacteristically, he finished the book so quickly that it was ready for publication a mere 18 months later, in November 1859. From this point on, Darwin's role in conceiving the theory of evolution would overshadow Wallace's. But Wallace, who died in 1913, well after Darwin, certainly deserves equal credit. Today's scientists, if not the general public, are aware of the part he played in unraveling "the mystery of mysteries."

Only 1,250 copies of *Origin* were initially printed—a modest print run that failed to anticipate the furious demand for it. The books sold out almost at once and went through another six printings within a short time. As Darwin had expected, *Origin* caused a commotion from the very first day it went on sale. Scientists and general readers alike immediately took sides, supporting or condemning the work. *The Times,* England's most prestigious newspaper, gave *Origin* an extraordinarily favorable review by Thomas H. Huxley, a noted scientist in his own right. Huxley was so overwhelmed by Darwin's work that he was moved to fire off a letter full of praise to the author:

> I finished your book yesterday … no work on Natural History Science I have met with has made so great an impression on me & I do most heartily thank you for the great store of new views you have given me …. As for your doctrines I am prepared to go to the Stake if requisite …. I trust you will not allow yourself to be in any way disgusted or annoyed by the considerable abuse & misrepresentation which unless I greatly mistake is in store for you …. And as to the curs which will bark and yelp—you must recollect that some of your friends at any rate are endowed with an amount of combativeness which (though you have often & justly rebuked it) may stand you in good stead—I am sharpening up my claws and beak in readiness.

Abuse and Misrepresentation: The Storm Begins

Sharpening claws, if not the stake, would be called for. It was only a matter of time before the theory's detractors mounted a ferocious counterattack. But Huxley, who became known as Darwin's "bulldog" for his passionate advocacy of the theory of evolution, was braced for the onslaught. When the British Association for the Advancement of Science met in June 1860, Darwin's opponents launched the opening salvo, heaping scorn on the ideas embodied in *Origin.* The Archbishop of Oxford, Samuel Wilberforce, ridiculed the whole notion of evolution and asked Huxley whether he was descended from an ape on his grandmother's side or on his grandfather's side. Huxley was more than happy to take the bait and rose to Darwin's defense. According to one version of the events, he declared, "I would rather be the offspring of two apes than be a man and afraid to face the truth."

All accounts of the tumultuous gathering agree that Huxley crushed Wilberforce in the debate, defending evolution as the best explanation yet advanced for species diversity. No one was convinced, however, who wasn't already partial to evolution to begin with. Among the holdouts was Admiral Fitzroy, the same Fitzroy who had captained the *Beagle.* An extremely devout man, he waved a Bible in the air, declaring that it was the unimpeachable authority and not "the viper" who had been aboard his ship. What made the whole furor even more astonishing was that nowhere in Darwin's book had he ever discussed the origin of man.

Although Darwin had been prepared for a storm, he didn't seem to expect the passions that *Origin* would stir up. He expressed his astonishment at the controversy. "I see no good reason why the views given in this volume should shock the religious feelings of anyone," he wrote. In fact, Darwin was anxious to avoid getting into any theological or sociological debates about his work. A reserved man who tended to shun publicity, he preferred to be left alone to continue his research and his writing. However, recognizing that he might have offended many of his readers, he hastened to assure them that he had no intention of dismantling religious belief. On the contrary, he sought to declare his own faith:

> There is grandeur in this view of life, with its several powers, having been originally breathed by the Creator into a few forms or into one; and that … from so simple a beginning endless forms most beautiful and most wonderful have been, and are being evolved.

Darwin's attempt to put God back into the great drama of nature, while sincere, did nothing to dampen the mounting protests.

Even if Darwin wasn't willing to engage himself in the trenches, other scientists were pleased to take his place. Of all the theory's defenders, Huxley proved the one who was most influential in getting scientists and the public to accept it. As a biologist, zoologist, and paleontologist, Huxley had the credentials—and credibility—to make his voice heard over the roar of the crowd. Huxley may have been "Darwin's Bulldog," but he wasn't slavish about his devotion. When he found flaws or problems in the theory, he wasn't afraid to let Darwin know about it.

What Darwin had done by putting out his theory was call into question two of the most cherished doctrines of nineteenth-century thought: the uniqueness of man and the traditional view of cosmic history. One of the central themes of Christianity was the special relationship of man and God in history. Evolutionary theory, however, seemed to suggest that natural processes weren't directed by divine providence at all, but rather were an unforeseen consequence of the blind action of natural selection. If all evolution was merely the cumulative result of a long series of "mistakes," what did that say about the purpose of life? What did that say about the existence of a God who watched over his creation? Christians weren't the only ones who took offense—so did some Marxists (but not Marx himself). "If it could be proved that the whole universe had been produced by such Selection," wrote the irascible playwright and essayist George Bernard Shaw, "only fools and rascals could bear to live."

Darwin's theories of evolution challenged nineteenth-century Christian belief on at least four grounds:

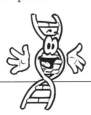

A Natural Selection

To put the contribution of Darwin and Wallace into context, it should be noted that the role of chromosomes in heredity wasn't discovered until 1903, and DNA wasn't identified until 1943.

➤ By emphasizing that species changed, evolutionary theories apparently destroyed ancient notions of the great chain of being, in which all living organisms had their proper place in a fixed, immutable order.

➤ By emphasizing that species changed over time, evolutionary theories called into question the validity of Genesis as the true account of creation.

➤ Evolutionary theory undermined the central place that human beings occupied in the world, in a manner similar to the Copernican revolution, which demonstrated that the Earth was not the center of the universe.

➤ Natural selection, based on random variations and the apparently wasteful cruelty of the selection process, undermined the belief in a moral God.

Finding Flaws in Darwin's Theory

Many scientists lined up against Darwin, too, seizing many of the technical problems, gaps, and inconsistencies in Darwin's work to demolish the entire edifice that he had constructed. Among the problems they pointed out, and which Darwin acknowledged, was the problem of assessing the actual extent of variability in nature. Nor had Darwin determined how new favorable traits could take root in the wild, when there was a likelihood of crossbreeding with other species.

Was there was any limit to variation in nature? Darwin wasn't sure. Just how extravagant was nature with its variations? Again, Darwin didn't know. Human breeders of hybrid plants, domesticated animals, and racehorses, for instance, could produce a great deal of variation—and in a fairly short time. He doubted that, left to its own devices, nature could do the same. In addition, it took Darwin time to appreciate how different physical and biological environments—known today as ecological niches—could create positive opportunities for new forms to exploit and encourage divergence and adaptive variations.

These weren't the only difficulties that Darwin confronted. What was the mechanism for transmitting traits from one generation to another? Lacking workable theories of inheritance and variation, Darwin could only fall back on what could be observed of variation in domesticated animals and plants. And if that wasn't enough, Darwin had yet to determine what the cause of variation was in the first place. All he could do was lay aside the question and accept the fact of variation as a starting point.

Other scientists, including Darwin's former geology professor at Cambridge, Adam Sedgwick, objected to the theory on the grounds that natural selection was amoral and materialist, and so was degrading to humanity's spiritual aspirations. In a letter to Darwin, he began by acknowledging that organic development was unquestionably a "fact of history" that could be "denied by no one of common sense." But having said that, Sedgwick took issue with Darwin's idea of natural selection:

> There is a moral or metaphysical part of nature as well as a physical. A man who denies this is deep in the mire of folly. Tis the crown & glory of organic science

115

that it *does* thro' final cause, link material to moral …. You have ignored this link; &, if I do not mistake your meaning, you have done your best in one or two pregnant cases to break it. Were it possible (which thank God it is not) to break it, humanity in my mind, would suffer a damage that might brutalize it—& sink the human race into a lower grade of degradation than any into which it has fallen since its written records tell us of its history.

Who Are They?

Alexander Agassiz (1835–1910), the son of the famous scientist Louis Agassiz, was an American zoologist specializing in jellyfish and corals. He served as curator of the Museum of Comparative Zoology at Harvard from 1873 to 1885.

Sedgwick was echoing many of the criticisms leveled at evolutionary theory by devout Christians: There seemed no room in Darwin's scheme for God or morality.

Darwin's theory also came under attack from another well-regarded scientist, Louis Agassiz, who was unable to reconcile himself to a theory that did not recognize a divine plan for nature. Agassiz was one of the last adherents of natural theology—a belief that science and religion could be embraced without any contradiction, which had once inspired countless scientists, including Darwin himself. However, by the time of publication of *The Origin of Species,* natural theology had largely run out of steam because it failed to offer any plausible explanation for how life had evolved. Much to Agassiz's dismay, most of his students—including his own son, Alexander, a well-known naturalist—became evolutionists, although not necessarily Darwinians.

The Problem of Better-Endowed Males

Origin was Darwin's crowning achievement. For the rest of his life, Darwin wrestled with the problems that he had been unable to solve in his ground-breaking book. In 1868, he published *The Variation of Animals and Plants Under Domestication* (1868), followed by *The Descent of Man* (1871). In writing the *Descent of Man,* Darwin took up a subject that he had paid scant attention to in *Origin,* which was how to explain the phenomenon of secondary sexual characteristics.

For instance, how could he account for traits in a number of organisms that seemed to have no survival value at all, such as the brilliant colors of a peacock's tail? Darwin made a stab at an answer in *Descent:* "I am convinced … that many structures which now appear to us useless, will hereafter be proved to be useful and will therefore come within the range of natural selection." All the same, he acknowledged that the theory could do with improvement. Nevertheless, he went on to say this:

> I did not formerly consider sufficiently the existence of structures, which as far as we can at present judge, are nether beneficial nor injurious, and this I believe to be one of the greatest oversights yet detected in my work.

That suggested to Darwin that another form of natural selection must be involved:

> It is clear that these characteristics are the result of sexual and not of ordinary selection, since unarmed, unornamented, or unattractive males would succeed equally well in the battle for life and leaving a numerous progeny, but for the presence of better endowed males.

He conceded that various "unimportant characters" were acquired through sexual selection.

Where did this leave Darwin's theory? Darwin was convinced that he was on solid ground in proposing natural selection as the principal instrument for evolution. But natural selection, at least as originally proposed, didn't explain other aspects of development, such as the problem of *secondary sexual characteristics*. And while his theory could largely account for the bodily development of man, it didn't say anything about the mental or spiritual differences that separated humans from other primates. On the other hand, Darwin was inclined to believe that the minds of humans and animals, while considerable, were "certainly one of degree and not of kind." That Darwin would think this way is hardly surprising; if all life is related by shared common ancestry, then it makes sense that the higher primates—humans included—would share certain mental abilities as well. The question of whether humans and animals also had a similar emotional life led to his next book, *The Expression of the Emotions in Man and Animals* (1872).

What Does It Mean?

Secondary sexual characteristics are attributes not necessarily related to reproduction that distinguish one sex from another—larger breasts in women, for example, or deeper voices in men. Primary sexual characteristics, by contrast, are attributes directly related to reproduction.

For all intents and purposes, though, *Descent* marks the end of Darwin's efforts to grapple with the major problems of evolution. Instead, he focused his remaining energy more narrowly on botany. In the late 1870s, he published *The Effects of Cross Fertilization in the Vegetable Kingdom* and *The Different Forms of Flowers on Plants of the Same Species*. Writing to a friend on his 73rd birthday, Feb. 12, 1882, he wrote, "My course is nearly run." He died two months later, on April 19th, and was buried in

Westminster Abbey in London—an honor only given to English subjects who have made significant contributions to the scientific, cultural, and political life of the country.

Although Darwin had struggled to remain aloof from the conflict that raged over his theory, he couldn't completely escape it—not even in death. Shortly after he died, newspaper stories began to appear claiming that he had undergone a deathbed conversion and repudiated his theory. Darwin's daughter was quick to dismiss the reports. "I was present at his deathbed .... He never recanted any of his scientific views, either then or earlier."

Even as opponents continued to debate and denounce Darwinian theory, many prominent scientists had moved beyond the controversy. They turned their attention to the challenge of working out many of the problems and inconsistencies that Darwin had failed to resolve. But, like Darwin, they were hamstrung by their ignorance of the mechanism of reproduction and inheritance. A real understanding of the way in which evolution actually works in nature would have to wait for a monk in Austria to take an interest in peapods.

The Least You Need to Know

➤ Alfred Russell Wallace derived a theory of evolution that, in almost every major respect, was similar to Darwin's.

➤ Both Darwin's and Wallace's theories of evolution are based on the idea of natural selection as the instrument of variation in species and the creation of new species.

➤ The presentation of the theory of evolution, first in a scientific gathering in London in 1858 and later in Darwin's major work, *The Origin of Species,* triggered an immediate uproar.

➤ Many devout Christians opposed the theory of evolution on the grounds that it seemed to suggest that there was no divine purpose in nature, a view that they shared with many scientists of the day, including Darwin's former geology professor, Adam Sedgwick.

➤ Other critics seized the existing gaps, errors, and discrepancies in parts of the theory to call into question the validity of the whole thing. Most scientists, however, agreed that the theory was correct.

➤ Darwin freely admitted that he hadn't solved many of the problems raised by his theory (including an explanation for why secondary sexual characteristics existed in many species, even though they seemed to confer no particular survival benefit).

Part 3
Genes and Evolution

While Darwin's theory provided a satisfying explanation about how evolution worked, it was still riddled with holes—and no one was more aware of them than Darwin himself. Possibly the biggest hole of all was his ignorance of how traits such as eye color, body build, and temperament were transmitted from the parents to their off-spring. Intuitively, he had grasped why adaptive traits are preserved in nature from one generation to another, without, however, having understood the mechanism for inheritance.

The answer, however, lay within Darwin's grasp—literally. All he would have had to do was to open a letter sent to him by an Austrian monk named Gregor Mendel. Mendel, who had read Darwin's The Origin of Species, *wrote to Darwin about some ingenious experiments he'd been conducting with pea plants. The results of Mendel's studies in a monastery garden were to lay the foundation for modern genetics. But Darwin would never learn about them because he tossed the letter away, unopened. But as we discover in Part 3 of our story, genetics, far from undermining Darwinian theory, actually offered a scientific method of testing—and eventually proving—its basic principles. In that sense, we can say that Mendel rescued Darwin from his own mistakes.*

Chapter 11

The Man Who Saved Darwinism

In This Chapter

➤ The survival of the fittest

➤ The blood theory of heredity

➤ Gregor Mendel's experiments

➤ The principle of separation

➤ The principle of independent assortment

We've seen how Darwin's theory of evolution threw the world into a tailspin. Opposition to it came from many quarters, but, ironically, some of the biggest blows that it suffered didn't come from detractors, but from its supporters. That's because Darwinian theory was misunderstood and exploited by intellectuals with agendas of their own.

The result was the pernicious doctrine of survival of the fittest, which was meant to refer to the way in which humans struggled for money, power, and resources. Those who ended up the "winners" in this competition were obvious: They were the rich and powerful. So, it followed that the poor were the "losers." This was not what Darwin had meant at all by his theory of natural selection; what is known as social Darwinism represents a complete distortion of his views. Social Darwinism had some very dangerous consequences, as we'll see later in this book. But Darwinism wasn't just threatened by misinterpretation; it also faced some very serious difficulties on the scientific front.

The central problem confronting Darwin and other scientists who favored evolution was how to account for the mechanism that caused such staggering diversity in nature. And if Darwin was right in saying that the same mechanism responsible for the diversity also could eventually produce new species, the theory would somehow have to say what exactly that mechanism was.

We know that there are more than 2 million existing species of plants and animals, and some scientists suspect that as many as 10 million to 30 million more may still be out there, waiting discovery. Forget about the numbers for a moment. Think about how different in size, shape, and behavior these species are, with bacteria at one end of the scale, measuring less than one-thousandth of a millimeter in diameter, and the sequoias of Northern California, which tower 300 feet above the ground and weigh several thousand tons, at the other end. Or, think of how adaptable species are, with bacteria flourishing at temperatures close to the boiling point in the hot springs of Yellowstone National Park, and fungi and algae somehow managing to thrive in saline pools at –9° F, not to mention the strange worm-like creatures discovered in dark ocean depths at thousands of feet below the sea surface, subsisting in a poisonous sulfurous atmosphere. How could this astonishing versatility and adaptability be explained?

It took an Austrian monk with a scientific bent and an intuitive genius to come up with an explanation. Very simply, he was able to demonstrate the way in which heredity is transmitted. His name is Gregor Mendel, and his theory forms the basis of our understanding of genetics. It was only by grasping the way in which inheritance is passed along—from innovative experiments with simple pea plants—that the "mystery of mysteries," as Darwin called it, could finally be penetrated. We can even go so far as to say that Mendel's discovery of genetics saved Darwinism from the same dustbin into which so many previous theories of evolution had been discarded.

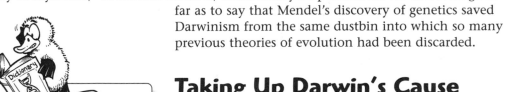

What Does It Mean?

Ethology is the study of animal behavior, mainly in the wild. It was developed in Europe (and the term is used mostly by European biologists) and is based on the idea that most animal behavior is intuitive or "hard-wired" in the animal's nervous system rather than learned.

Taking Up Darwin's Cause

After Darwin's death two men—Darwin's "bulldog" Thomas Huxley and the philosopher Herbert Spencer—spearheaded the movement championing the naturalist's theory of evolution. Huxley's most famous contribution, *Evidence on Man's Place in Nature*, published only five years after Darwin's *The Origin of Species*, offered a comprehensive review of what was known at the time about primate and human paleontology and *ethology*—animal behavior. More importantly, Huxley's was the first attempt to apply evolution explicitly to the human race, a subject that Darwin had avoided.

For his part, Herbert Spencer, a younger contemporary of Darwin, seized on Darwin's theory to advance his own, sometimes dangerous, ideas. All organic matter, Spencer wrote, originates in a unified state, with individual characteristics gradually developing through evolution. In 1860, a year after *Origin,* he wrote an ambitious book, *A System of Synthetic Philosophy,* in which he put forward a system of philosophy based on evolution that would embrace and integrate all existing fields of knowledge. (This was one of the first attempts to do so.) He was quick to embrace Darwin's theory of natural selection, but in a way that totally distorted what Darwin was trying to say. Spencer advocated a philosophy called *social Darwinism,* which was intended to apply Darwinian theory to human societies. (We'll get into the subject of social Darwinism later and see what belief in the doctrine ultimately led to. Without going into any detail, the short answer is, nothing good!)

Actually, it was Spencer to whom we owe the slogan "survival of the fittest." Human progress, he argued, resulted from the triumph of more advanced individuals and cultures over inferior individuals and cultures. As a result, it was assumed that if you had wealth and power, you were "fit"; if you were poor, you were condemned as "unfit." Darwin was deeply disturbed by this distortion of his theory. Nonetheless, in depicting a ruthless dog-eat-dog world, in which nature rewarded the rich and powerful and doomed the poor and unsuccessful, Spencer was doing the theory of evolution a grave disservice.

Who Are They?

Herbert Spencer (1820–1903) was a British social philosopher, considered one of the first sociologists and one of the nineteenth century's great intellectuals. He is recognized in particular for trying to systematize all knowledge within the framework of modern science—especially in terms of evolution. However, his reputation has suffered since his death as a result of his advocacy of social Darwinism, a distorted attempt to apply Darwinian principles to human society.

Matter and Motion: The Riddle of Reproduction

Darwinian theory, however, was stuck—and, for all the efforts of its supporters like Huxley and Spencer, it was never going to advance without an understanding of the mechanism of inheritance. For hundreds of years before Darwin, scientific thinkers had been struggling to penetrate the mystery of heredity, which was one of the most puzzling phenomena of nature. After all, sex cells, which are the vehicles for transmitting inheritance are invisible to the human eye. (It wasn't until early in the seventeenth century, with the invention of microscopes by Antoni van Leeuwenhoek, that it become possible for scientists to actually see and study the sex cells.)

Not that there wasn't some glimmer of an understanding of how heredity worked—almost by necessity, this came through agriculture. The ancient Babylonians, for instance, learned that pollen from a male date palm tree must be applied to the pistils of a female tree to produce fruit. Aristotle, the brilliant fourth century B.C.E. Greek philosopher, thought that, although both male and female parents contributed different traits to their offspring, the female supplied what he called "the matter," while the male provided "the motion." According to Hindu texts compiled between 100 and 300 C.E. in India, the role of the female in reproduction was like that of the field, and the male's role was like that of the seed. New bodies were formed "by the united operation of the seed and the field." (In reality, of course, both parents transmit their genetic inheritance equally.)

Then in 1694, a Dutch botanist named Rudolph Jacob Camerarius showed that the same reproductive process that the Babylonians had observed in date palms applied equally to corn. In experiments conducted about a century later, Carl Linnaeus (the great taxonomist that we talked about earlier) and another botanist, Gottlieb Kölreuter, crossed varieties and species of plants to see what the offspring would look like. On the whole, they discovered that the hybrids appeared to be a blend of the parents, although, in some characteristics, the offspring were closer to the male parent and, in others, were closer to the female.

Even more curious misconceptions about how heredity worked came into fashion. For instance, some people held to the belief of prepotency, a theory that gave more weight to some parents than others when it came to how their children would turn out. Some individuals, according to this misconceived idea, "impress" their heredities on their offspring more effectively than others. "Maternal impressions," for instance, imply that events experienced by a pregnant female would be reflected more in the constitution of the unborn child than "paternal impressions" would be.

Who Are They?

Josef Gottlieb Kölreuter (1733–1806) was a German botanist who became best known for his studies of plant fertilization. He began by crossing tobacco plants but then extended his experiments to many other plants. Later he crossed hybrids with each other in experiments that foreshadowed the pioneering work of Gregor Mendel.

New Blood, Pure Blood, and Bad Blood

When so many experiments on plant hybrids in the 1800s seemed to show that hybrids were average or intermediate between the parents, it confirmed the widely held belief in a theory called blood heredity. Although the blood theory of heredity is a myth, it still permeates our vocabulary today.

Think of the terms *half-blood, new blood, bad blood,* and *blue blood.* Biologists of the nineteenth century who espoused this theory didn't mean that heredity was literally transmitted by blood. Instead, they maintained that a parent transmits to each child all its characteristics and that the inheritance that each child receives is a blend of the hereditary contributions of its parents, grandparents, great-grandparents, and so on. If you had one short parent and one tall parent, the thinking went, you would turn out to be of medium height. And your children would be of medium height, too, unless the other parent was uncommonly tall or short.

Even Darwin was initially a believer in blood heredity. He proposed that the fundamental units of inheritance were minute particles called gemmules, which were produced in every part of the body and passed to the sex organs, where they were incorporated into the sperm and ova. By this logic, the original cells from which the embryo developed reflected the "condition" of the parents at the time of conception. In this way, Darwin suggested, characteristics acquired during the parents' lifetime could be transmitted to their children, a notion more associated with Lamarck than with Darwin.

Obviously, the idea of blood heredity was riddled with flaws. For starters, how could the same parents transmit different "bloods" to different children? How often have you seen children of the same parents who are so different from one another—intellectually, physically, and temperamentally—that you're surprised to find that they're related? While it's true that many parents do have offspring who seem to be an average of both, there are some glaring exceptions. And those exceptions caused several scientists to reevaluate the whole notion of blood heredity.

Scientists focused in on the finding that some characteristics in families reappeared after having disappeared for many generations. A trait as innocuous as red hair or as insidious as Huntington's disease (a fatal brain and nerve disorder) can appear in one generation, disappear for four or more generations, and then suddenly crop up in a child born a century later. Such a strange and disturbing phenomenon could even occur when a breed was crossed only once. This kind of delayed action of inheritance didn't fit with the idea of blood heredity or a "blending" of characteristics in the offspring supplied by the parents.

Lamarckism didn't go to its grave without a struggle, though. People seemed to like the idea that characteristics acquired during your lifetime could be passed along to your children. It took a biologist named August Weismann to prove that modification of individuals by environmental influences, for instance, had no effect on descendants. He did this in a series of famous experiments in the late 1890s by amputating tails in generations of mice. The next generation—and those that followed—had tails the same length as their parents before their amputation. This isn't to say that environmental influences cannot have long-lasting genetic effects. X-rays, some drugs, and powerful chemicals, for instance, can mutate genes; as a result, a parent may be capable of passing such modified genes to the offspring. (Most mutations are not harmful, though, nor do they usually have any discernible effects.)

Although Darwin later distanced himself from the idea of blood heredity, he never discovered how inheritance was actually passed on. Ironically, he came close to the truth twice. A regular habitué of the London pigeon racing clubs, he had observed that, while the offspring of two different stocks of pigeons were the same type, cross-breeding of the birds over several generations produced such a variety of pigeons that hardly any two of them were alike. It was just that Darwin never took the next step to inquire why this should be—why did traits that might not have been seen in either the original pigeon stock or the original "foreign" pigeons crossbred with them show up several generations later among their descendants?

The second time that Darwin had the opportunity to reexamine his thoughts about inheritance came with the arrival of a letter, this time from the very person who was to revolutionize the understanding of heredity: Gregor Mendel. Mendel had written a paper on his experiments with hybrid peas, disproving once and for all the theory of blood heredity: Offspring are not a blend or an average of their parents. (We'll get to Mendel very shortly.) Mendel was aware of Darwin's work—he had read *Origin* and made many notes in its margins—and was convinced that his own work might have a good deal of bearing on the subject of evolution. Had Darwin read Mendel's paper, he might have appreciated just how significant it was and tried to incorporate them into his own theory, thereby solving many of the most difficult problems still bedeviling his work. But Darwin never opened the letter.

Who Are They?

August Weismann (1834–1914) was a German biologist who is known for his gene-plasm theory of heredity. This theory postulates the idea that some special hereditary substance (which turned out to be DNA) constituted the only organic continuity between one generation and the next.

A Monk and His Pea Plants

Born in 1822, Gregor Johann Mendel became interested in natural science at an early age—a result of his childhood experiences roaming through his father's orchard and farm. Ironically, when he took the qualifying examination for certification as a teacher, he not only failed, but he received one of his lowest marks in biology! Fortunately for science, the setback didn't discourage him. Even after being ordained as a priest and joining a monastery in Brno, Czech Republic (as it is known today), he continued to teach himself science and actively participated in meetings of the city's Natural Science Society. From all accounts, it appears that Mendel began his famous experiments on pea plants before he read *Origin* sometime in the 1860s.

The Tall and the Short, the Wrinkled and the Smooth

Mendel had two things going for him: He was a genius and he was lucky. He was lucky in his choice of what plant to study, and he was a genius because he was able to perceive what no one before him had seen—that the products of inheritance (offspring) differ from the mechanism of heredity (genes). Peas, it turned out, were like many garden plants in that they have a pure lineage—they have been kept isolated from crossbreeding for so long that, within any given line of descent, every plant is identical or purebred. (Or, you can think of them as inbred.) But pea plants within different lines are distinct. Lines of pea plants exhibit many traits, including height, pea shape and texture, and the color of its flowers and seeds, even in the position of the flowers on the stems. In addition, peas are both male and female and can fertilize themselves.

Given all these distinctions, Mendel had a ready-made basis for comparison. What would happen, Mendel wondered, when he crossbred pea plants of different lines? What kind of traits would their offspring demonstrate after one or several generations? And then what would happen if these hybrid offspring were mated with one another? In theory, anyway, these experiments had an elegant simplicity about them, but, surprisingly, in the 10,000 years of agricultural breeding, no one had ever tried them.

Mendel began by taking male sex cells—pollen—from a pure line and then using it to fertilize the eggs of plants of another line. Each inbred family that he chose was selected for one particular trait—color, for example. Pea plants in different lines were either yellow or green. When he crossed a plant with a pure yellow line with another with a pure green line, he was surprised to see that all of the offspring were yellow. Obviously, this result meant that there was no validity to the idea that progeny were a blend or an "average" of their parents. Otherwise, you would expect offspring green-yellow in color. Peas looked like one parent and not the other.

Mendel now decided to cross the offspring to see what would happen. The result of this experiment was also astonishing, only in a different way: Now three out of every four plants were yellow, and one out of every four was green. Some factor had been responsible for the reappearance of green in the second generation. That suggested to Mendel that the green agent must be contained in a plant whose seeds were yellow. This agent of heredity we now call the *gene*. And here was the secret of Mendel's genius: The agent of inheritance, Mendel saw, had an existence separate from that of its vehicle, the plant.

No matter how often Mendel repeated these experiments, the ratios remained the same: In the second generation, it was alway   s thre to one green. The agents of inheritance, Mendel deduced, must come in two forms, or *alleles*, to cause this effect. Body cells contain a pair of alleles for each trait, while each reproductive cell (in plants, the pollen or egg cell) receives just a single one. They combine in different way   s

What Does It Mean?

A **gene** is a unit of inheritance, composed of genetic material that is responsible for the inheritance of a particular characteristic or set of characteristics.

a yellow and a green, two greens. A recessive allele must be present in a double copy to show itself. That means that all green peas have two green alleles.

What Does It Mean?

Alleles are different forms of the same gene. Although both alleles are carried in the genetic material of each parent, the parent contributes only *one* allele to the offspring. Alleles are either dominant or recessive. If one is dominant, only that one will express itself in the next generation. If each parent contributes recessive alleles to the offspring, however, the recessive allele will be the one to express itself.

The yellow peas in Mendel's first generation descended from parents of different colors, each bearing a single copy of the yellow allele, matched with a single copy of the green. When Mendel crossed these hybrid plants—producing a third generation—a quarter of the plants had two yellow alleles, a quarter had two greens, and a half had one of each allele. Because the yellows are dominant and the greens are recessive, all plants with yellow alleles would produce yellow plants. But when two recessive green alleles were paired, the plants were green. This meant that three quarters of the plants were yellow, and one quarter were green. That gives the famous Mendelian ratio of 3:1.

Traits Lost and Found

Once Mendel had seen how color was inherited, he went on to carry out the same experiment on other traits of the pea plant, such as height, seed color, seed shape, the position of the flowers on the stem and the form of the pods. To his delight, the same ratio applied once again. Each of these traits was governed by dominant and recessive alleles, and the result each time was the same reassuring three-to-one rule. These experiments also showed how a trait "lost" or hidden for generations could reappear. Every time the allele for the lost trait was paired with a dominant allele, it never manifested itself. When, by random chance, it was matched with another recessive allele, the trait emerged again. Such rare recessive alleles, Mendel discovered, are almost invariably hidden unless both parents carry the recessive alleles.

The pea plant experiments seemed to suggest that the mechanism of inheritance, once found, operated in a relatively straightforward way. But nature has more tricks up her sleeve than Mendel realized: When he then turned up to the study of another plant, the hawkweed, he failed to come up with similarly satisfying results that would corroborate this theory. This was because he didn't realize that hawkweeds had the frustrating (for him, if not for the plant) ability to reproduce without sex—a phenomenon known as *parthogenesis*. As a result, there was no possibility of producing any genetic variation.

In his experiments disproving the blood theory of heredity, Mendel showed the following:

➤ Heredity is transmitted through elementary units of information (now called genes) that do not blend, but remain separate.

➤ Parents transmit only half of the genes that they have to each child, and different children receive different sets of genes.

➤ The occurrence of constant variation in the characteristics of species and their descendants is a result of the occurrence of paired elementary units of heredity or genes.

➤ Although brothers and sisters receive their heredities from the same parents, they do not receive the same heredities (except when they are identical twins).

➤ In sexually reproducing organisms, humans included, every individual has a unique hereditary endowment.

What Does It Mean?

Parthogenesis refers to the ability of certain plants and insects to reproduce without sex. Some female aphids are able to reproduce without any males, for example.

By showing that these elementary units obey simple statistical laws, Mendel established that, in the reproductive cells of the hybrids, half the cells are responsible for transmitting one parental unit to the offspring, and half are responsible for transmitting the other unit. In other words, the alternative characteristics, conveyed by each unit or gene, are kept apart in reproduction. There is no blending. This principle is known as Mendel's first law, or the principle of segregation.

What happens over several generations of mixing and matching of these units? Well, sooner or later, as Mendel recognized, you will have all possible combinations of dominant and recessive alleles showing up in the offspring. Mendel had seven pairs of different characteristics (color, height, pod shape, and so on) to work with in pea plants. That meant that he was starting with seven pairs or 14 units of heredity (one from each parent.) So, he was able to observe how, over successive generations, the seven pairs of characteristics recombined at random.

This gave rise to his second law, the principle of independent assortment. It states that genes combine randomly. This principle, however, applies only to genes that are transmitted in different "linkage groups." That is, you need to have crossbreeding—green and yellow pea plants reproducing with each other, for example. The same principle doesn't affect reproduction within a purebred line. Green pea plants breeding with other green pea plants will produce only more green pea plants.

Mendel had done something else important, too. He had shown that life wasn't based on liquids, such as blood, as had previously been thought, but on particles or units of information. (Today we could say that genes operate by means of a digital language.) Genes can be recovered unchanged by time, even though individuals carrying them continue to die. We can even think of species—ourselves included—as convenient instruments for passing along genes. Because information is passed on through these genes, long-lost characteristics may eventually emerge in a distant descendant. In some cases, the genes themselves can be older than the species that carries them!

Mendel Lost and Found

Mendel presented his theory and a description of his experiments at two meetings of the Natural Science Society of Brunn (Brno) in the winter of 1865. He declared that none of his predecessors had done as much as he in penetrating the secrets of "plant hybridization." Mendel was right not to be modest. While many experiments had been done on plant hybrids throughout the 1800s, none of Mendel's predecessors grasped the significance of the data that they had gathered. A year later, in 1866, the transcript of the meetings was published under the unexciting title of "Experiments with Plant Hybrids." It seems to have had no effect whatever—not on the scientists who had attended his talks or scientists anywhere else, even though the paper reached major libraries in Europe and America. While Mendel corresponded with eminent botanists of his time, apprising them of his work, they didn't seem to appreciate what he had accomplished, either.

After publishing one last paper in 1869, Mendel ceased all his scientific investigation, discouraged as much by the failure of his hawkweed experiments as by the indifferent reception to his work from the scientific community. Although he continued to have an interest in botany, bee culture, and meteorology, he gave more attention to his duties as a monk. After being elected to the abbot of his monastery, he soon became bogged down in tedious administrative obligations.

Mendel's work had come at a time before anyone truly realized the essential role heredity played in producing evolutionary change. Like so many other scientists, he was ahead of his time. His contribution to the understanding of heredity (and evolution) wasn't recognized until 1900—six years after his death—when other scientists resumed research where Mendel left off, crossing a variety of plants and animals to see whether the Austrian monk's results held up. They did.

By formulating the essential criteria for how heredity works based on his original experimental data, Mendel had solved the problem that had eluded Darwin and other evolutionists like Huxley. Darwin's critical question—how varieties could be transformed into other species—now had an answer in Mendelian theory. In 1900, three other European botanists, Carl Erich Correns, Erich Tschermak von Seysenegg, and

Hugo De Vries, independently obtained results similar to Mendel's. But they weren't aware that they had gone over the same territory that Mendel had already covered. It was only when they went back and searched the scientific literature that they found, to their surprise—and probably dismay—that both the experimental data and the general theory for heredity had been published 34 years previously. Fame had finally come to Mendel, but, unfortunately, too late for him to enjoy.

The rediscovery of Mendel's work marks the beginning of the history of genetics, in which biologists in many countries conducted experiments to confirm and expand upon the theories worked out in a secluded monastery in Austria. As we'll see, Mendelian theory would serve as the basis of a rapidly developing science which was to have astonishing influence on the understanding of evolution, development, physiology, biochemistry, medicine, agriculture, and social science.

Who Are They?

Hugo Marie De Vries (1848–1935) was a Dutch botanist who independently corroborated Mendel's work on heredity. De Vries was responsible for introducing the concept of genetic mutation into evolutionary theory.

The Least You Need to Know

➤ Herbert Spencer misapplied Darwin's theory of natural selection to human society to promote his idea of survival of the fittest, implying that nature intended the rich and powerful to be the "winners" of the struggle for resources.

➤ Darwinist theory still confronted a major problem: It could not account for the mechanism of heredity—how inheritance is passed from one generation to the next.

➤ Even in Darwin's time, many scientists believed that offspring were a blend or an "average" of both their parents.

➤ The savior of Darwinism was an Austrian monk named Gregor Mendel, who showed by his experiments on pea plants that heredity operates by elementary units or particles of information, with units being contributed equally by both parents.

➤ By showing that recessive characteristics are manifested because of a pairing of two recessive alleles, Mendel was able to explain why traits hidden for generations can turn up again.

➤ Mendel's major work, carried out in the 1860s, laid the foundation for modern genetics.

Chapter 12

The ABCs of Natural Selection

In This Chapter

➤ Finding the agents of heredity

➤ Genetic variety as the key to natural selection

➤ The role of alleles in variation

➤ How genetic variation contributes to survival

➤ The blindness of national selection

For all his achievements, Mendel had never identified the agent of heredity. It fell to scientists who rediscovered his work to understand that this agent is a gene, which is carried on chromosomes in the nucleus of every cell in the body. (These cells are called somatic cells. *Somatic* refers to the body.) Further scientific investigation showed that diseases were genetic in nature.

As biologists learned more about genetics, they also began to recognize the role genes played in evolution. They were able to fill in many of the blanks in Darwinian theory, honing in on the crucial issue as to how exactly natural selection works. Natural selection, as Darwin recognized, was the primary mechanism of evolution. But it was only after scientists had a handle on how genes mutate to endow certain individuals, whatever their species, with traits that make them more "fit"—allowing them to adapt to new environments.

Adaptive traits are those that confer an advantage, however slight, on an individual and increase the individual's chances of reproductive success. That means that the genetic variants—the positive adaptive traits—possessed by the individual are likely to be more frequent in the offspring of the next generation. As we explore the machinery of evolution, you'll learn how natural selection acts on species as diverse as finches and moths, sorting out the winners and disposing of the losers, all based on their genetic inheritance.

Brown Urine and Colorless Skin: The Discovery of Genes

In formulating his laws of heredity, Mendel had postulated the existence of elementary units of inheritance. But what these units were composed of or where exactly in the cells they were located was unknown to him. We now know that these units are actually genes, that they are composed of *DNA* (deoxyribonucleic acid) and that they are carried by *chromosomes* in the nucleus of every cell. Even though Mendel was operating to some extent in the dark and carried out his experiments only with varieties of peas, his two fundamental laws of inheritance—the principle of segregation and the principle of independent assortment—turned out to apply equally well to the inheritance of many kinds of characteristics in practically all organisms. Just to keep things straight in your mind, here is what these two laws said:

➤ **The principle of segregation.** Each parent contributes one unit (gene) of inheritance to its offspring. The units are kept apart or segregated. The offspring is not a "blend" of the parents' genetic inheritance.

➤ **The principle of independent assortment.** Genes combine randomly, meaning that, over time, they will pair together in practically every conceivable combination. This random assortment can cause traits to re-emerge in progeny after having disappeared over the course of several generations.

What Does It Mean?

Chromosomes are microscopic units in each cell containing DNA, the hereditary material that governs the development and characteristics of an organism. In most organisms, chromosomes are complex structures with DNA molecules arrayed in a linear pattern (with the exception of bacteria, where they are found in a circular pattern.)

In 1902, having reclaimed Mendelian theory from the obscurity into which it had fallen, scientists demonstrated these two laws in poultry and in mice. A year later, albinism became the first human trait shown to be caused by a recessive gene. (You might recall that the eighteenth-century German scientist Louis de Maupertius had remarked on the same phenomenon when reports reached him of albinism running in Senegalese families.) Specifically, albinism, which is characterized by a loss of skin pigmentation, was caused by a pair of recessive genes. Or, to put it another way, the dominant alleles (forms of a gene) in humans code for pigmentation.

However, when two recessive alleles are paired in an offspring—which happens every so often because of the random combinations predicted in Mendel's second law—the result is an aberrant condition that results in a loss of pigmentation. In the decades since, geneticists have discovered many such anomalies caused by recessive genes, including several serious inherited diseases.

Linking Genes to Disease

The first time a disease was linked to a genetic cause came in the early 1900s. In 1903, a British physician, Sir Archibald Garrod, treated a 3-month-old boy with a condition known as alkaptonuria, which causes urine to turn a deep reddish brown. The condition is typified by the abnormal accumulation of a substance called alkapton in the body. Over the next several years, Garrod interviewed the families of 39 patients with the illness. Interestingly enough, none of the parents of the affected children, but all the couples were first cousins.

Before Mendel, it would have been difficult to explain why the parents, while related to one another, escaped the affliction that affected their children. But, by applying Mendel's laws, Garrod was able to unravel the mystery. (Actually, Garrod initially suspected a bacterial infection, but once he'd ruled that out, he realized that a genetic link was a probable cause.) The condition, he found, was due to the pairing of recessive genes that impede the normal action of an *enzyme* that breaks down alkapton into carbon and water. All 39 families carried the recessive genes for alkaptonuria; if they had married people they were not related to, the chances of having children with the condition would have been considerably diminished. That's because the recessive gene for alkaptonuria is relatively rare in the general population.

What Does It Mean?

An **enzyme** is a specialized organic substance, composed of amino acids, that act as catalysts to regulate the speed of chemical reactions in organisms. About 700 enzymes have been identified. The name, suggested by German physiologist Wilhelm Kuhne comes from the Greek phrase *en zymc* ("in leaven").

The Search for the Stuff of Life: Chromosomes, DNA, and the Genome

The pace of scientific progress picked up considerably in the twentieth century. Scientists conducting cellular studies using ordinary light microscopes found that chromosomes are the carriers of genes. (As early as 1848, biologists had observed that cell nuclei resolve themselves into small rod-like bodies during mitosis, the process by which the nucleus of the cell divides and replicates. Because these structures were able to absorb certain dyes they were named chromosomes, meaning "colored bodies.") In the late 1930s, using experimental and observational evidence, American geneticist Theodosius Dobzhansky showed that, although mutations do occur, they usually don't have a significant influence over a species' reproductive success. Too many mutations in a species, he found, will make them incapable of reproduction, so they will be unable to pass along the mutated genes to the next generation.

What Does It Mean?

Nucleotides are the chemical compounds that make up DNA, the basic genetic material containing the hereditary information in all cells of the body. Each nucleotide is composed of a phosphate, a sugar called deoxyribose, and any one of four nitrogen-containing bases: adenine, thymine, guanine, and cytosine.

Then, in 1943, the Canadian scientist Oswald Avery proved that DNA carries genetic information. He proposed that DNA might actually be the stuff of which the gene was made. It took only a few more years before DNA was largely accepted as *the* genetic molecule. What still wasn't known was how the DNA molecule was structured. In 1950, yet another American scientist, Erwin Chargaff, succeeded in determining the composition of DNA. In 1953, James Watson, an American scientist, and Francis Crick, a British scientist, discovered the structure of the DNA as a double helix—two linked chains of chemical compounds known as *nucleotides*. The groundbreaking work, which spurred the genetic advances that have revolutionized science and society, won the two the Nobel Prize for Medicine.

Genes, we say, are what are responsible for the traits and behaviors of an organism. So the genetic discoveries of the twentieth century handed scientists an invaluable arsenal of new tools to explore how inheritance works that had never been possible before. Keep in mind, though, that when we talk about "traits" in terms of Mendelian inheritance, we are actually using a convenient abstraction. Traits come bundled together, and many genes, directly or indirectly, may be involved in their expression.

Who Are They?

Arhibald Edward Garrod (1857–1936) was a British physician who was knighted for his work on inherited diseases. He was one of the first scientists to apply Mendelian theory to the study of human diseases.

In 2000, scientists completed the sequencing of the human genome. The genome is commonly described as a genetic "blueprint" or "map." However, it may be more usefully thought of as a "toolbox" full of implements that make us who and what we are. To the scientists' surprise, they discovered that humans are composed of about 30,000 genes, a humbling comedown from the 100,000 estimate that most geneticists had previously assumed. (To put this in perspective, the simple roundworm has about 9,000 genes.)

That finding suggested that genes do double, triple, and quadruple duty, alone and in various combinations: A single gene may affect many traits. Just one gene in Drosophila flies (which have been widely studied because of their primitive genetic structure) affects the color of the eyes, the reproductive system in both males and females, and the longevity of both sexes. In humans, a single defective gene not only can cause many diseases, but it also can produce a variety of symptoms that, on initial examination, may appear to have nothing in common.

The Link Between Genes and Evolution

Evolution can be defined as genetic change through time. More specifically, we can think of evolution as change in gene frequencies of a population over time. Evolution occurs because different individuals who are genetically different tend to leave different numbers of offspring to grow, mature, and reproduce in the subsequent generation. As this process is repeated from generation to generation, a change begins to take place in the genetic composition of the population.

The mechanism for this change is natural selection. But evolution and natural selection are not synonymous. *Evolution* refers to changes in the genetic structure of a population, however these changes come about. *Natural selection* specifies one very important way in which genetic change can occur. Of course, genetic change can come about only through reproduction. (There are a few exceptions, as we have noted, such as exposure to large doses of radiation, which can affect an individual's genes during his lifetime.) We've seen why Mendel's theories work so well in explaining how genetic inheritance operates. But what value does Mendel's theory have when it comes to understanding how evolution works?

That old cliché about variety being the spice of life is only half true. In fact, variety, it turns out, *is* life. Without variety, there would be no evolution. Remember what Darwin said in *The Origin of the Species:*

> [I]t may be asked, how is it that varieties, which I have called incipient species, become ultimately converted into good and distinct species, which in most cases obviously differ from each other far more than do the varieties of the same species? How do those groups of species ... which differ from each other more than do the species of the same genus, arise? All these results ... follow inevitably from the struggle for life.

Then Darwin offers the answer:

> Owing to this struggle for life, any variation, however slight and from whatever cause proceeding, if it be in any degree profitable to an individual of any species, in its infinitely complex relations to other organic beings and to

137

external nature, will tend to the preservation of that individual, and will generally be inherited by its offspring.

Note the words *any variation*. The only qualification is that it has to be a variation
that confers an advantage that helps an individual adapt to its environment. So, if we
agree that variation is the mechanism that makes evolution possible, the next question is, how do you achieve the necessary variation? And what kind of variation are
we talking about, exactly?

The Spice of Life

That is where Mendel comes in. The kind of variation we are referring to is genetic
variation. How does this variation occur? Remember what Darwin wrote about evolution being based on an accumulation of "mistakes." Those mistakes take the form of
genetic mutations. Simply put, mutation is a change in a gene. As we have said before, these mutations may be positive or harmful, or may have no effect whatsoever.
These mutations may occur as a result of an exposure to an external influence in an
individual's lifetime—radiation, toxic chemicals, and so on. More common are mutations that occur because of errors in the genetic copying system that transfers hereditary information from parents to offspring. These changes, good, bad, or indifferent,
are the source of new genetic variation. Natural selection operates on this variation.

Variation doesn't occur at the same rate or to the same degree. For evolution to take
place, there must be mechanisms to create genetic variation, increase that variation,
and decrease it. Considerable variation is seen in natural populations. Birds have
much less variation (about 15 percent) than insects, while mammals and reptiles fall
somewhere in between.

Alleles—the different versions of the same gene—play a crucial role in genetic variation. For example, humans can have A, B, or O alleles that determine one aspect of
their blood type. Most animals, including humans, are diploid, meaning that they
have two alleles for every gene at every location or locus, one that they inherited
from their mother and one that they inherited from their father. *Locus* refers specifically to the location of a gene on a chromosome. Humans can be AA, AB, AO, BB,
BO, or OO at the blood group locus. That covers all possible combinations of alleles
for this particular trait.

The link between genes (in terms of their variant allele forms) and evolution can be
observed in moths and explains why some types of moths are still in existence and
others have never come into being. In moths, one allele results in a moth that has a
tail, and another allele codes for a moth without a tail. In addition, a gene determines whether the wing is brightly colored or dark. So, we have four possible types of
moths: brightly colored moths with tails, brightly colored moths without tails, dark
moths with tails, and dark moths without tails. Genetically, all four of these types
can be produced. That is, there's no genetic barrier to any one of these four types. In

fact, all four types can be produced in a lab. But only two types of moths are actually found in the wild: brightly colored moths with tails and dark moths without tails. Why should this be so? What does nature have against brightly colored moths without tails and dark moths with tails? The answer is natural selection.

Over time, moths, like all existing species, have developed strategies for survival. One strategy is called *mimicry*. Imitation, it turns out, isn't simply a sincere form of flattery; it's also—at least in the wild—a means to live to see another day. Moths with bright wings and tails resemble another insect species that tastes terrible to birds, which are their natural predators. If not for the deception, the birds would probably find the moths a delicious treat. In the tropics, there's even a phenomenon known as mimicry rings, in which many species of butterflies, moths, and sometimes other insects—some palatable to predators, others not—have developed similar patterns and colors on their wings. So, the birds, confused as to which tastes good and which disgusts them, leave them all alone. A brightly colored moth *with* a wing, on the other hand, doesn't resemble the insect that birds have learned to avoid. That would make them easy pickings, and they wouldn't survive for very long. So, this form of moth doesn't appear in the wild at all.

What about dark moths with tails? They pursue another strategy for survival that is called cryptic, which refers to camouflage. The dark wings allow the moths to blend in with tree trunks, where they gather during the day. The strategy gets even more elaborate still: Some dark moths have wings that duplicate the patterns of lichens or mosses that grow on the bark. In either case, they are practically invisible to hungry birds and other predators. If dark moths with tails were produced in nature as they can be in the lab, they wouldn't last very long because the wing would call attention to itself and the camouflage wouldn't work. So, we see that, even though the genetic inheritance of these moths allows for four forms, only two of them confer an advantage to the insects in contending with their natural environment. Are both strategies—mimic and cryptic—equally successful? Do both brightly colored winged moths and darkly colored wingless moths survive in the same numbers? If you read on, you will learn the answer.

Who Gets to Contribute Their Genetic Legacy

We can define natural selection as the reproduction and survival of individuals carrying favorable, inherited traits. It involves differences in the relative contributions of various genotypes to the next generation. A *genotype* is a term that refers to the overall genetic makeup of the individual. In other words, individuals are chipping in their particular genetic inheritance—to varying degrees—to the

What Does It Mean?

Genotype refers to the underlying genetic makeup of a trait or to the overall genetic makeup of the individual. A genotype can also be described as a genetic variant.

139

population as a whole. From the standpoint of a gene, individual members of any species are only vehicles capable of getting them from one generation to the next. However, a gene is not the agent of selection (because its success depends on the organism's other genes as well). The agent of selection is the individual organism, which either reproduces or fails to reproduce.

Which genotypes get to contribute most to a population depends on a number of factors:

➤ Various environmental factors, including competitors, predators, parasites, and climactic changes

➤ Differences in fertility and fecundity among genotypes

➤ Differences in frequency of reproduction among genotypes

➤ Differences in viability/longevity among genotypes

We can say that increases in certain genotypes in a population are ultimately due to different rates of survival and different rates of reproduction. This means that some types of organisms within a given population leave more offspring than others. Over time, the frequency of the more prolific type of genotypes will increase. So, another way of defining natural selection is the difference in reproductive capability. Natural selection can also be thought of as a kind of censor, sifting through the available genotypes and removing unfit variants as they arise via mutation.

Depleting Diversity

When alleles that are deleterious are weeded out, genetic variation is depleted. Sometimes, though rarely, an allele is favorable enough that it becomes rooted or "fixed" in a population. After all, evolution usually takes place over long periods of time. (There are exceptions, as we noted in our first chapter, such as bacteria or viruses that undergo rapid evolution—for example, in developing resistance to drugs.) But, in the short run, nature is more interested in maintaining stability. Stability in a species, in this sense, requires genetic stability or equilibrium as well so that the overall genetic diversity of a population is not being changed significantly.

Not all deleterious alleles are weeded out, though, so they continue to be passed to offspring. Let's take the example of sickle-cell anemia in human populations. The populations affected by this illness live in tropical countries, especially in Africa, where malaria is often rampant. Red blood cells are either normal shaped or shaped like a sickle. If the blood cell is sickle shaped, it cannot carry normal levels of oxygen. If you have inherited two alleles for sickle cells from your parents, you will develop anemia, a painful and debilitating illness.

You might think that if natural selection were doing its job correctly, this would be the very type of allele that it would remove from the gene pool. But it turns out that if you were living in a malaria ridden country and inherited only one allele for sickle cell and one normal allele, you would actually enjoy a degree of resistance to the disease. That's because the shape of sickle cells makes it harder for the plasmodia—malaria causing agents—to penetrate the blood cell. By the same token, if you had two normal alleles, you would be more likely to suffer from malaria. As a result, the sickle cell allele offers an advantage. People with a single sickle cell allele and one normal one will pass on both alleles to the next generation without having the illness. Thus, neither sickle-cell nor normal blood cell allele can be eliminated from the gene pool. Not surprisingly, the sickle-cell allele is at its highest frequency in regions of Africa where malaria is most pervasive.

Selecting for and Selecting Against

Let's go back to the moths. Although both brightly colored and darkly colored moths have adapted to elude their predators—mainly birds—the dark moths are the clear winners. They have a higher reproductive success for the simple reason that there are a lot more of them. A cryptic strategy turns out to be superior to a strategy of mimicry. However ingenious the mimics are, they still fall victim to predators at a greater rate than their darker cousins. With more brightly colored moths culled out of the gene pool, the number of alleles coding for bright colors declines in frequency as well. We can say that brightly colored moths are being selected against. Alleles will change in frequency depending on the different reproductive rates of the individuals whom they inhabit. Organisms with high reproductive success will inevitably contribute more of their alleles to a population, and those with limited reproductive success won't contribute even close to as many. So, certain alleles will be selected for and others will be selected against.

As in Darwin's time, the Galapagos continues to be an unparalleled natural laboratory in which to study the operation of natural selection. Fifteen species of finch live on the archipelago. One of them, *Geospiza fortis,* has a relatively small beak and feeds on smaller seeds from a particular plant. Another species, *G. Magnirostris,* has a larger beak and feeds on larger seeds. Both species owe their survival and their long-term prospects to the availability of food. But in 1977, the environment changed because of a drought. With diminished rainfall, fewer seeds of each kind were produced. Understandably, that put the seed-eating finches under a great deal of stress. The *G. fortis* population soon ran through all the small seeds they could find, but, without enough to sustain them, most of the finches starved. Their population plunged from about 1,200 to less than 200. However, there still remained a sufficient quantity of large seeds to feed the larger-beaked finches. The larger birds continued to reproduce and had offspring that also had large beaks. The result was an increase in the proportion of large-beaked birds in the population in the next generation and a sharp drop-off of the smaller-beaked birds.

But was this a true genetic change? To find out, a researcher named Peter Grant crossed finches with various beak sizes and succeeded in showing that a finch's beak size was, in fact, influenced by its parent's genes. So, here is an example of natural selection in action. The environmental influence did not directly cause an increase in larger-beaked finches at the expense of the smaller ones—a view that is essentially Lamarckian. Instead, an environmental factor (drought) brought about a change in the availability of the food source. One species (the larger-beaked finches) had a genetic variant (bill size) that made it more suited—or "fit"—to adapt to the environmental change, and the other species (the smaller-beaked finches) had a genetic variant (bill size) that made it less fit to adapt. The fitter species, having adapted more successfully, was thus in a better position to reproduce and bring into the world birds that would also have larger bills. So, we can say that the larger beaked finches were selected for and the smaller-beaked finches were selected against.

Now let's suppose that the environment changes once again. Imagine a scenario in which conditions now produce an abundance of smaller seeds while causing larger seeds to become relatively scarce. Over the years since the drought, the population of larger-beaked finches has grown considerably (because they are better adapted for the existing environment), while the population of the smaller-beaked finches has diminished. But without a sufficient number of large seeds, many of the larger-beaked finches won't get enough to eat and will starve. Meanwhile, the smaller-beaked finches will have an abundance of food because there are fewer of them compared to the amount of food available. Given enough time and no further dramatic environmental change, the smaller-beaked finches will reproduce more, and their population will gradually expand even as the number of larger-beaked finches declines.

Natural Selection Has Limits

Bear in mind that genetic variation itself isn't caused by natural selection. The variation already exists in the species. Natural selection simply distinguishes among variants (or genotypes) that already do exist, promoting some and rejecting others. There are limits, however, to what natural selection can do. Variation is not endless in nature. That means that there are limits to the possible adaptive solutions available to a species.

Turtles, for example, do not develop shells made out of steel, even though, as an adaptive solution, this wouldn't be a bad idea. After all, a steel shell would offer more protection from two-ton cars that account for the daily slaughter of turtles on the highway. But no genetic variation is available in turtles to code for metal shells. This places constraints on what natural selection can do to help the turtles to adapt to an environmental hazard that earlier generations of turtles never had to face.

Natural Selection Is Blind

Natural selection doesn't operate with any foresight or long-range objective. It is a mechanism that simply allows organisms to adapt to their current environment. What the future will bring, natural selection has no idea and could care less about. Structures or behaviors do not evolve because they might be useful in the future. An organism adapts to its environment at each stage of its evolution. As the environment changes, as our example from the Galapagos demonstrates, new traits may be selected for or against. When large changes in populations occur (which is what happened to the Galapagos finches), they are the result of cumulative natural selection.

Natural selection works in the following way:

➤ All species have a degree of genetic variation—some more, some less.

➤ Genetic variation occurs because of mutations.

➤ Faced with dramatic environmental change, species are forced to adapt or are eliminated.

➤ Individuals who have genetic variants that make them better able to adapt to the new environment are more likely to survive and have offspring.

➤ The offspring of the fittest individuals will carry the same genetic variants to one degree or another; over time, the population of these better-adapted individuals will grow.

➤ Natural selection does not induce genetic variation; it merely distinguishes among different variants and selects those variants that confer an advantage for adaptation.

➤ Genetic variants are not limitless.

Although the way in which natural selection works is fairly well established, many questions still puzzled geneticists. For instance, how do complex traits evolve? Just because a trait is currently useful for one function doesn't invariably mean that it always served the same purpose at an earlier stage of evolution. Traits can evolve for one purpose and later prove useful for another. Bird feathers, for instance, are thought to have evolved not for flying, but as insulation and possibly as a way to trap insects. Later, proto-birds may have picked up the ability to glide when they leapt from tree to tree, until eventually the feathers were co-opted to keep the birds airborne.

The reverse case is also seen. Penguins, for example, evolved from birds that once flew. Now penguins use their wings only for swimming. A trait evolved for its current usefulness is an adaptation; one that evolved for another type of use is called an exaptation.

The Least You Need to Know

➤ Without genetic variation, natural selection couldn't take place.

➤ Genetic variation occurs because of mutations in the genes.

➤ Some individuals have traits that confer an advantage in particular environments; these individuals will be more capable of adapting and will consequently have more reproductive success.

➤ Natural selection does not induce genetic change; natural selection selects for existing favorable traits and selects against unfavorable traits—that is, those that do not help an individual adapt to an environment.

➤ Traits developed for one purpose—bird feathers intended originally for insulation, for instance—may be subsequently used for another purpose, such as flying.

Chapter 13

Who Is the Fittest One of All?

In This Chapter

➤ The selfishness of natural selection

➤ Reciprocal altruism—is it really altruism?

➤ Survival of the fittest—what it actually means

➤ Why common genes (and rare ones) stick around from one generation to the next

➤ The foundation of population genetics

In one sense, this is a chapter about why words don't always mean what we commonly think they mean. What does a term like *selfish behavior* mean to you? If you know someone who's selfish, you probably would want to steer clear of him. Yet when biologists talk about selfishness, they mean something entirely different.

Or consider the TV series *Survivor*. It's possible that if you've been on Mars, you might not know about the spectacular hit, but you would be one of the few. In *Survivor*, a group of people face a series of contrived challenges and, in each installment, vote one another off the island until only one is left to claim the million-dollar prize. *Survivor* evokes a kind of Darwinian scenario, a simulated survival of the fittest. As it turns out, though, the simulated contest on *Survivor* has nothing at all to do with what Darwin meant by *survival of the fittest*. (If it did, ratings would probably be even higher.) In fact, even the word *fit* means something different to evolutionary biologists than it does to nonscientists.

These linguistic distinctions aren't academic, either. They go to the heart of what evolutionary theory is all about. Think of natural selection as a high-stakes game in which the winners are organisms that not only survive but that also are capable of contributing more of their genetic legacy to the next generation. But how do you prove as much? It was one thing for scientists and mathematicians to theorize and speculate about the extent of genetic variation in populations (yes, mathematicians play a key role in this story, too); it was quite another to be able to actually see and study this variation in the cells.

That didn't stop the development of a new science called population genetics in the first decades of the twentieth century. Even if they had no knowledge of what took place in the cells, scientists could always theorize about how genetic change might occur over successive generations. Still, they were unable to account for just how much variation might occur. Nor could they prove whether it might be possible to detect evidence of evolution in progress. With the advent of new technologies in the middle of the twentieth century, however, it became possible to actually answer some of these questions.

However, as always in science, finding answers to old questions only raises new ones. Scientists began to look at natural selection from a new angle. Yes, they agreed, selection is a crucial force in evolution, but maybe other factors were involved in evolution that needed to be studied as well. Of course, first they had to find out what those factors were.

The Survival of the Fittest: What Does It Really Mean?

Individuals aren't particularly loyal to other members of their species—and why should they be? For the most part, they are competitors with one another. Altruists would be selected against in a given population because the selfish individuals would reap the benefits without having to pay the price. Biologists use the words *selfish* and *altruistic* in a way that is different from the rest of us, though. They do not mean to imply any moral judgment. In biological parlance, *selfish* simply means behaving in a way to maximize one's own fitness, while *altruistic* means behaving in a way to increase another individual's fitness at the expense of one's own. This is instinctive behavior, not behavior that is pursued consciously.

This isn't to say that cooperation doesn't occur in nature, although many types of behaviors appear altruistic. Ants and bees, for example, live and work in a cooperative fashion. Biologists, however, believe that cooperation or helping fellow organisms is often the most selfish strategy for an animal. They call this phenomenon reciprocal altruism. *Vampire bats* that have found a meal will regurgitate some of the blood into a famished partner's mouth. (This is a phenomenon that few readers will probably want to investigate further.) However, if a bat cheats by accepting blood when it needs it but fails to donate it when his partner needs it, the stingy cheater will face social ostracism and be abandoned by its partner. So, biologists say, the bats are not really helping each other altruistically, but rather are practicing behavior that benefits them in return for their co-operation. The practice is mutually beneficial. This is why the term *reciprocal altruism* is used.

Surviving Is Only Half the Battle

What exactly does reproductive success consist of? Here's where the whole idea of survival of the fittest needs to be reexamined. Reproductive success, according to biologists, is based on fitness. The phrase *survival of the fittest* is often used synonymously with *natural selection*. But survival is only one component of selection—and possibly not always the most important one, at that. In some species, for instance, of the males that make it to reproductive age, only a few ever mate. So, survival is by no means a guarantee that males will find mates or reproduce. The ability to survive and the ability to attract mates are not the same thing. You can observe as much in a singles bar any night of the week. Survival, we say, is a necessary condition to be fit, but it isn't sufficient.

Another misconception that needs to be cleared up relates to what exactly *fitness* means. It's understandable why the word is usually associated with exercise, health clubs, nutritious diets, and so on. In other words, being fit, to most people, means physically fit. But that's not how evolutionary biologists use the term. Fitness, in their definition, is the average reproductive output of a class of genetic variants in a *gene pool*. That is, the more individuals in a population that contribute their genes to the next generation the more "fit" that population is considered. Fit does not necessarily mean biggest, fastest, or strongest.

What Does It Mean?

Vampire bats are bats that feed on the blood of other animals. The three species of vampire bats are the common vampire bat, the white-winged vampire bat, and the hairy-legged vampire bat, all of which are found in the Southern Hemisphere.

What Does It Mean?

The **gene pool** is the sum total of the genes carried by the individual members of the population.

Defining Reproductive Fitness

Darwinian fitness refers only to reproductive fitness. Robust health, vigor, and mental acuity may all contribute to being reproductively fit, but they are not necessarily essential to it. For example, mules can be strong and resilient, and yet, as far as Darwinian fitness goes, they are big zeros because they are sterile. They have no chance of reproductive success at all. So, you can begin to see why, in viewing natural selection, evolutionary biologists today place less emphasis on studying survival than on reproductive success. This isn't to say that survival isn't important. After all, if individuals don't survive to reproductive age, they won't have any reproductive success whatsoever.

Reproductive success, then, comes down to who gets to make the most substantial contribution of their genetic inheritance to the gene pool. The contribution, though, isn't simply a single optimal genotype—a variant that confers an advantage—but rather an entire array of genetic variants that collectively enhance the ability of a population to survive. Less fit genotypes are less able to contribute toward the next generation, so they consequently are less represented among the offspring.

If we look at fitness in this way, as a capacity that has adaptive (or selective) value, it throws an entirely different light on the Darwinian description of natural selection as the survival of the fittest. While the word *fitness* may call to mind Stairmasters and workouts, the words *survival of the fittest* tend to evoke scenes from the Nature Channel where fast-moving predators rip out the hearts of wildebeest and zebras. Actually, this isn't what Darwin meant at all. Yes, he said, there was a struggle to survive. But "struggle" didn't necessarily mean contention, strife, or combat. Struggle didn't have to lead to a life-or-death gladiatorial contest with one competitor triumphing to fight another day and the other sprawled out bleeding in the dust. Natural selection doesn't require such a dire outcome to be effective.

Natural selection favors traits or behaviors that increase a genotype's inclusive fitness. Closely related organisms share many of the same alleles. In many species, including humans, siblings share an average of at least 50 percent of their alleles. (Alleles, you may recall, are different forms of the same gene.) So, it makes sense to help close relatives to reproduce—an example of reciprocal altruism—because it will increase the likelihood that an organism's own alleles are better represented in the gene pool. If you think of it as a game, the winners would be those who got to pass along the most gene variants to the next generation. The contribution doesn't have to come only directly from you, either, as long as a significant portion of your genetic legacy gets into the gene pool of the offspring. Relatives are especially valuable in this respect in highly inbred species. (That's because so many of them share so many of the same genes; thus, the more of them that can reproduce, the more their genes will flourish and spread in succeeding generations.)

In an even more dramatic example of reciprocal altruism, sometimes organisms will forgo reproducing at all, to help their relatives reproduce. For instance, ants as well as other social insects, including some species of bees, have sterile castes dedicated to the service of the queen. Their only purpose is to help her reproduce. By contributing their efforts in this way, the sterile workers are reproducing by proxy. That is, their genetic legacy is getting into the next generation, even though, individually, they are only indirectly contributing to it.

Just how many offspring will carry the favorable genetic variants in the next generation—who will then be able to reproduce, in turn—depends on a great many factors: the ability to compete, find mates, avoid predators, forage, and survive harsh weather conditions, among others. Each generation has the same evolutionary obligation: to survive and reproduce in order to leave relatively greater numbers of their offspring to reproduce in the subsequent generation. Biologists say that fitness has an adaptive (or selective) value because it refers to the relative ability of an organism to contribute its genetic inheritance to subsequent generations.

Some aspects of fitness need to be emphasized:

➤ Fitness equals reproductive success.

➤ Fitness is a characteristic of genotypes, with a class of individuals having many genetic characteristics in common.

➤ Fitness is concerned with the average number of offspring produced by an individual of a particular genotype. An individual who leaves more offspring is more fit.

➤ Fitness does not imply health or vigor, except as such attributes contribute to reproductive success.

➤ Fitness can be considered only in the context of a particular environment.

➤ Biologists refer to two types of fitness—direct and indirect.

➤ Direct fitness is a measure of how many alleles, on average, a genotype contributes to the next generation's gene pool by reproducing.

➤ Indirect fitness is a measure of how many alleles identical to the individual's own the individual helps to enter the gene pool. (An individual doesn't need to contribute his own alleles if he helps a relative reproduce.)

➤ Inclusive fitness is direct fitness plus indirect fitness.

Diving into the Gene Pool

Every population has a gene pool. The gene pool of a population (called a Mendelian population by biologists) is not a static thing; it changes over time. The genes of the individuals of a generation that's alive today come from a sample of the genes of the previous generation. If individuals of an existing generation reproduce, their genes will pass into the gene pool of the next generations. From your own experience, you know that individuals who live and work close together are more likely to meet and to mate than people who live far apart. That was why, until recently, people who lived in different countries or continents rarely met, let alone married or had children. That resulted in a human gene pool that was fragmented into smaller gene pools of races and populations distributed widely in many parts of the world. Just as other species run into natural barriers, such as mountains and oceans that prevent them from ranging far and wide as they might otherwise do, humans, too, confront natural limits.

Although we are all of the same species—more than 99 percent of our genetic makeup is shared—that isn't the same as saying that we all tap into the same gene pool. Geography isn't the only factor that keeps gene pools apart; there are also linguistic, religious, social, economic, and educational barriers that further divide the gene pools. Often these gene pools overlap or subdivide. The smallest subdivision is referred to as an *isolate* or *panmictic unit;* it consists of a relatively limited number of persons (or animals or plants) that can be considered as potential mates. However, these subdivisions are not so sharply defined that you can say precisely where one gene pool subdivision begins and another one ends. Nonetheless, these categories are considered biologically meaningful because they make each subdivision easier to examine and study.

A biological species is defined in terms of its sexual reproducing organisms. Any species in which all its members can reproduce only with one another, and with no other species, is known as the most inclusive Mendelian population. Humans, then, are an inclusive Mendelian population. No gene exchange occurs between humans and any other related species, like such primates as monkeys or apes.

Human gene pools are never pure—all gene pools have the capacity to absorb genes from even far-flung human gene pools. There may be no known instance of an Eskimo and a Melanesian having children together, but genetic transmission between the Eskimo and Melanesian gene pools takes place anyway. That's because Eskimo and Melanesian genes are spread by populations that come into contact with both groups. As a result, if a favorable genetic change occurs anywhere in the world, it stands a chance of spreading throughout the entire human population. Some biologists speculate that this may have been how genetic changes transformed the ancestral prehuman species into us—*Homo sapiens.* Conversely, any genetic damage resulting in mutations (from exposure to radiation, for example) to one group has the capacity to spread into populations that are far away.

A **panmictic unit** refers to a population in which random breeding occurs. It is derived from the Greek meaning "mingling" or "mating."

Tracking Down Genes Through the Generations

In general, you'd expect that genes that provide superior reproductive efficiency would increase in frequencies from generation to generation, while those genes that do not would decline in number. Biologists wondered how and if these frequencies would change. Many scientists assumed that the more common versions of genes would increase in frequency simply because there were already so many of them in the population to begin with.

Surprisingly, though, rigorous studies carried out in the twentieth century showed that sexual reproduction is actually a conservative force that tends to maintain the genetic status quo in a population. Variety is all very well and good in terms of evolution, but at any given time, populations tend to remain relatively stable. In other words, every species isn't undergoing evolution at every time. It turns out that it doesn't matter whether an allele is common or rare; it will ordinarily remain as frequent (or as rare) in successive generations.

If a gene frequency is 1 percent in a population, it tends to stay at 1 percent indefinitely unless some force acts to change it. There are two basic ways in which to impose a change. You can do it in the laboratory or on the farm—you can introduce a gene (or knock one out) in a mouse or produce a resistant strain of grain never before seen in nature. That's artificial selection. Otherwise, the most powerful force for changing gene frequencies is our old friend, natural selection.

The Hardy-Weinberg Rules: Is Evolution Taking Place or Not?

In 1908, British mathematician Godfrey Harold Hardy and German physician Wilhelm Weinberg independently demonstrated that the frequency of alleles, common or not, does not change over time. They proposed what is known as the Hardy-Weinberg Rule, a set of algebraic formulas describing how the proportion of different genes can remain the same in a large population. The rule has a predictive value. Scientists can compare a population's expected genotypic frequencies to its actual genotypic frequencies to determine whether the population is maintaining the same ratio, or equilibrium, of genotypes over time.

Hardy and Weinberg stated that this equilibrium remains the same in a population under a number of conditions—among them, individuals must mate randomly and without regard for appearance (*phenotype*), no new alleles can be produced by mutation, and the population has to remain high. The reason why it matters that people choose their partners randomly is that, otherwise, certain traits will tend to be favored over others. This could have the effect of altering the frequency of genes or alleles in a population. The population needs to remain high because there are simply more alleles available, so their frequency can be more accurately measured.

If a population meets the conditions set by the Hardy-Weinberg Rule, it will maintain the same proportions of different genes over time. That means that the genetic makeup of the population never changes. Rare genes will never disappear, and common genes will remain just as common as before. The Hardy-Weinberg Rule has proven useful because it allows biologists to see whether evolution

What Does It Mean?

A **phenotype** refers to the appearance of an individual produced by the interaction of genes and environment. It comes from the Greek meaning "show" and "type."

is taking place. If the frequency of genes in a population is in equilibrium, scientists can safely conclude that evolution isn't taking place. However, if existing genes are replaced by new genes or genes with more selective advantage, the evidence would suggest that natural selection may be at work and some members of the population are producing more or stronger or healthier offspring. As we've pointed out, natural selection isn't the only mechanism for change. Changes can also be caused by other factors, including genetic mutation, migration, a decrease in population size, or a phenomenon called genetic drift. All these factors occur naturally over time.

Who Are They?

Sewall Wright (1889–1988) was an American geneticist who originated the mathematical theory of evolution, in which he stated that mathematical chance, in addition to mutation and natural selection, affects evolutionary change.

Who Are They?

John Burdon Sanderson Haldane (1892–1964) was a British geneticist who applied mathematics to determining the rate of genetic changes in human populations. His work in human genetics inspired him to study hemophilia and color blindness to discover the rates of mutations that caused these disorders.

Population Genetics

By the 1920s, when British geneticist John B.S. Haldane did some of his most significant work, scientists had determined that genes were located on the chromosomes (even if they weren't sure exactly where), that genes sometimes mutated spontaneously, and that the rate of mutation could be increased by radiation and certain chemicals. It was also known that many traits were partially controlled by many genes, each of which made incremental contributions to the development of the trait. At this point, though, no one knew what the genes were made of—that wouldn't come until the 1940s—nor did they have any idea how they worked.

In spite of their ignorance of how evolution worked on a microscopic level, several scientists and mathematicians continued to build on Weinberg and Hardy's work. Two mathematicians—Sewall Wright of the United States and Sir Ronald Fisher of England—and John B.S. Haldane helped to further develop mathematical theories of evolution. They were trying to figure out which organisms are best adapted to their environment so that they pass on their genes to their progeny. The collective work of Wright, Fisher, Haldane, and others laid the groundwork for a new branch of science known as population genetics, a field of biology that attempts to measure and explain the levels of genetic variation in populations.

For a long time, though, population genetics was completely theoretical. Gathering the data needed to test the theories was nearly impossible. Until the development of molecular biology, which required technological advances not available to the founders of population genetics, estimates of genetic variability could only be inferred from levels of morphological differences in populations. (Morphological differences refer to differences in the bodily structure of organisms.) In the 1960s, however, geneticists developed several techniques that allowed them to study specimens of DNA or proteins by observing how they respond in the presence of a slight electric charge. For the first time these scientists had a way of quantitatively determining the genetic change that occurs in the formation of a new species. One of the most important of these techniques is called molecular electrophoresis, in which DNA or proteins are separated and purified in an electrical field. Such techniques proved that populations varied extensively at the molecular level.

However, scientists also found that not all of this variation had any discernible benefit. Then in 1968, Japanese geneticist Motoo Kimura proposed that much of this variation—at least, on the molecular level—arises not from natural selection at all, but from chance mutations that have nothing to do with an organism's fitness or reproductive efficiency. Other geneticists and evolutionary biologists, however, disputed his ideas. What is beyond dispute, however, is that several mechanisms aside from natural selection play a role in evolution—as we'll find out in our next chapter.

The Least You Need to Know

➤ Natural selection is selfish: Organisms do not act altruistically toward others because they would imperil their own chances to reproduce while enhancing the chances of the donors to reproduce.

➤ Fitness in Darwinian terms does not refer to health or vigor, but rather to reproductive success.

➤ Although competition is perpetually taking place in nature—over resources, mates, territory, and so on—it does not necessarily imply a life-or-death struggle.

➤ There are limits to the spread of genotypes (genes governing a particular complement of traits) because of geographic, educational, linguistic, and other factors; nonetheless, any favorable or unfavorable mutation of a gene has the potential to spread to all humanity.

➤ The study of gene frequency in populations gave rise to a new science in the early decades of the twentieth century called population genetics.

➤ The development of advanced scientific technologies has allowed geneticists to examine cells on a molecular level so that, for the first time, they can see the genetic variation produced by evolutionary forces such as natural selection.

Chapter 14

The Natural Process of Loss

In This Chapter

➤ The mechanisms of evolution

➤ Natural vs. artificial selection

➤ Sexual selection: good gene and sexual runaway models

➤ Genetic drift

➤ How natural selection and genetic drift decrease genetic diversity

As we've seen, the principle of natural selection is based on several naturally occurring environmental and genetic phenomena. Every species is capable of geometric or exponential population growth. But growth cannot be allowed to run amok—that's when natural selection steps in to make certain that things don't get out of hand, often by means of some environmental disaster (flood, famine, disease, and so on) that forces the survivors to adapt or undergo what's known as "genetic death"—the destruction of certain genetic variants in a population.

An organism that doesn't leave any offspring—even if he lives a very long life—is committing genetic death by failing to pass on its genetic legacy. (An organism may escape this fate by helping close relatives pass on their genetic legacy, however—that way, the organism can ensure that the offspring will have some of its genes.) An individual organism's survival capacity is related to its genotype, or particular complement of genetic characteristics.

Survival, however, is important only insofar as the survivor is reproductively efficient so that it can pass along its genes. An organism that dies sooner than another has fewer opportunities to reproduce and, as a result, will have fewer offspring in its lifetime. By the same token, if an individual fails to find a mate to begin with, it won't have any reproductive success at all. That's why, as you'll see, finding a mate is so important in evolutionary terms. Sexual traits have developed that not only don't help an organism survive, but that may actually put the organism at risk.

By eliminating the ill-adapted—the unfit—from a population, natural selection is eliminating a lot of harmful genetic characteristics. That's good and bad news. The bad news is that natural selection is also making the population less genetically diverse. That also means that the population, especially if it's relatively small, has less protection against environmental disaster, especially disease. Moreover, through a process called genetic drift, more genetic diversity is being lost. So, if evolution was to continue, it had to develop offsetting mechanisms to add to genetic diversity. That's part of our story as well, a subject that we'll take up in Chapter 15, "Mutants, Shuffled Genes, and Mateless Maggots."

What Natural Selection Is—And Isn't

Selection is not a force in the sense that gravity or the strong nuclear force is, although biologists sometimes refer to it that way for the sake of convenience. So, it's wrong to say that the environment "pushes" or "pressures" a population to assume a more adapted state. Natural selection is limited; it will favor beneficial genetic changes whenever they appear by chance, but, as we said earlier, it does not cause genetic change to take place.

In fact, the potential for selection may be in place long before the occurrence of a favorable genetic variation to select. It's kind of an "all dressed up and no place to go" scenario. One of the problems that arises from referring to natural selection as a "force" is that it suggests that somehow selection has a mind of its own. This is not the case at all. Natural selection is simply a consequence of a fortuitous combination of events: a change in the environment and the existence of favorable genetic attributes that enhance an individual's chances for adapting to it and reproducing.

While we've described natural selection in general terms, scientists have broken it down into a number of different processes—sexual and artificial selection, for example—and studied how it takes place on a number of different levels, in closely related groups, populations, species, and so on. Scientists use several different criteria in making these distinctions.

Natural vs. Artificial Selection—What's the Difference?

Long before Darwin, farmers and breeders of horses, dogs, and pigeons, among others, practiced artificial selection. Breeders often attempt to increase some particular trait in a species by progressive selective breeding. A farmer, for instance, may hope to develop a crop with a greater yield, make a faster thoroughbred, or produce a cuter and cuddlier dog. To do this, the breeder will select the individuals with the desirable traits. He will then select only those offspring that express those traits and breed them. Eventually, the genes for these traits will become so frequent in the species that the traits will appear in almost every individual in subsequent generations. That accounts to a great extent for how we got from a wolf to a poodle.

In recent years, breeders now have another option: They can introduce genes that code for desired traits into the egg or embryo, custom-tailoring the type of organism that you want to see. (These techniques can be used on humans, too, offering the potential for designer babies.) Artificial selection, unlike natural selection, has a goal. Natural selection, as we've said, is blind, with no goal except for enhancing the chances of adaptation to a particular environment.

Sexual Selection: What's Love Got to Do with It?

One of the criticisms directed at Darwin's work was his failure to account for why, in many species, males develop prominent secondary sexual characteristics. This was a phenomenon that his theory of natural selection hadn't accounted for, as Darwin himself acknowledged. After all, as he remarked, a peacock's flamboyant tail didn't seem to do very much to help it survive. On the contrary, many of these traits would appear to be a liability from the standpoint of survival because they could alert a predator to the presence of the males that carry them. Nonetheless, the persistence of these sexual traits over long periods of time clearly indicates that they must serve some evolutionary purpose—why else would they be there? There were too many such manifestations to ignore—the brilliant coloring and patterns in male birds, the mating calls of frogs, and the flashes of fireflies. How then could natural selection favor these traits?

Remember what we said in the last chapter about the meaning of fitness. Natural selection has several components, of which survival is only one. Reproductive success matters as much or more. (Survival, of course, is the bottom line. If you die before adulthood, you're not about to pass along your treasured genetic legacy to the next generation.) Finding a mate is one of the most preoccupying activities in nature. That means that sexual attractiveness enjoys a high priority in natural selection. Sexual selection refers to factors that contribute to an organism's mating success. Even though certain traits may jeopardize an organism's survival, they can evolve because they enhance the organism's sexual attractiveness and, thus, its reproductive success. The value of the trait in enticing a mate outweighs its liability for survival. As far as natural selection is concerned, a male who lives a short time but produces many offspring is much more successful than a male who lives a long time but produces few offspring. Eventually, the genes from the short-lived males will dominate the gene pool of the species.

As a result, in many species, only a few males hold a virtual monopoly on the much larger number of females. These species are characterized by a phenomenon known as sexual dimorphism, in which two forms of the male occur. For instance, many species of male pheasants will develop two or three spurs on each leg that they use to defend their harems against male competitors. These spurs have a dual function in that they serve as defensive weapons and also indicate better genetic stock to discriminating females. Good fighting ability, in other words, equals good mating material. Some males have longer spurs, and other males have shorter ones. Males with longer spurs outlive their shorter-spurred rivals. Experiments have turned up the not-so-surprising result that females prefer the longer spurred males.

A Natural Selection

One type of sexual selection is known as intrasexual, in which males (and rarely females) compete through a display or a physical context for mates.

A Natural Selection

In epigamic sexual selection, females (or rarely males) search for mates displaying certain traits.

While the competition among males for mates can take the form of direct confrontation, in some species females choose who they prefer. In species in which females make the choice, males show off and engage in elaborate courtship rituals in hope of gaining favor. Usually the most outlandish displays produce the best results, a phenomenon that, among some human males, may take the form of lavish spending on five-star restaurants, expensive jewelry, and weekend Caribbean getaways.

The Good Genes Model: Explaining Bravado

Several theories have been advanced to account for how sexual selection operates. For instance, there is the good genes model, which states that extravagant display and bravado is an indication of male fitness. According to this model, a bright coloring in male birds is evidence of an absence of parasites. Selection for good genes can be seen in the stickleback fish. Stickleback males have red coloration on their sides. Studies showed that the intensity of the color was associated with both the relative number of parasites that the fish carried and the fish's sexual attractiveness. The fish that was reddest had the fewest parasites and attracted more females.

Runaway Sexual Selection Model

The good gene model has a competitor with a wonderfully evocative name. It's called the runaway sexual selection model. This model, proposed by Ronald Fisher, whom we referred to earlier, proposed that females may have an innate preference for some male trait even before it appears in a population. In this model, females would mate with males when, through natural selection, the trait eventually appeared. (It's like waiting around for Mr. Right to finally turn up.) In Fisher's scheme, the offspring of these parents would not only have the genes for the trait itself, but also would have the genes for the preference for the trait. If this process was allowed to continue indefinitely, the result would be a snowball effect. Scarcely any other kind of sexual traits would be allowed to get into the gene pool because of this feedback loop.

However, natural selection usually sees to it that a process like this eventually comes to a halt. As an example, let's suppose that a species of female birds prefer males with longer-than-average tail feathers. Males with the gene mutation for longer feathers will inevitably produce more offspring than the unfortunate males with short feathers. Over the next several generations, average tail length will increase—and increase more because, according to the runaway model, females have a hardwired preference for a longer-than-average tail, not a tail of some specific length. At a certain point the male tails will become so long that they will become a liability for survival. If the tail length makes it impossible to fly or puts a drag on the male's ability to move quickly, those birds will become easy prey, and then there will be many fewer male birds left to mate, regardless of female preference. In time, shorter tails will once again be in vogue and equilibrium will be established.

This isn't to suggest that there aren't many exotic male birds with rainbow-hued plumage or elongated feathers. In some cases, though, natural selection brokers a compromise: The males keep their feathers for as long as they are needed to attract a mate and then shed them after the breeding season. Fisher's model is meant only to explain how male secondary sexual characteristics and female preferences can go hand in hand.

We should also point out that the good gene and the sexual runaway models are not mutually exclusive. Millions of sexually dimorphic species exist on this planet, and there's little doubt that there is a great deal of variety in the forms of sexual selection that they exhibit.

A Natural Selection

In some species, individuals of one sex may monopolize individuals of the opposite sex during breeding season, creating a very competitive selective process. This practice is known as polygamy.

Genetic Drift: Why Panthers Are in Danger of Extinction

Toward the end of World War I, in the spring of 1918, a flu *epidemic* broke out in an army barracks in Kansas. People got sick for a week, took to bed, and then went on to make uneventful recoveries. Then, the following fall, the flu reappeared, but this time it was deadly. It swept around the world in a matter of months, felling young, healthy adults as well as children and the elderly. By the time it was over, 18 million people were dead, about half a million of them in America alone. Although no one is certain what made the virus so lethal, there is some evidence to suspect that the same flu virus, mild in the spring of 1918, returned in a slightly different but lethal form causing a *pandemic* in the fall. Somehow the virus had undergone a change in its genetic composition that had transformed it into a murderer. This phenomenon is known as genetic drift, a random change in allele frequencies due to mutation.

What Does It Mean?

An **epidemic** is an outbreak of a contagious disease affecting large numbers of people. It may last for a short time or persist for years. A **pandemic** is an epidemic on a global scale.

Alleles, as we've said earlier, are different forms of the same gene. As Mendel had shown, alleles can be dominant or recessive. When the offspring receives two identical copies of the same allele, they are said to be homozygous; when they receive one of each, they are said to be heterozygous for that gene.

Chance and the Genetic Advantage of Large Populations

Allele frequencies can change due to chance alone. Alleles can increase or decrease in frequency due to drift. There is no greater probability of the frequency increasing or decreasing. It is not uncommon for a small percentage of alleles to continually change frequency in a single direction for several generations. The process is as random as flipping a coin. You can keep coming up with heads seven or eight times in succession before you see a single tail. Genetic drift (as measured by the number of substitutions of alleles per generation) is independent of population size, although there is more variation in the rate of change of allele frequencies in smaller populations.

Large populations will have more alleles in the gene pool, so losing them slowly over time to drift won't make as much difference as it would in a small population, with fewer alleles to cycle through. But what is the result of the addition or subtraction of these alleles to the population? The answer is, it depends. Genetic drift may cause a population to lose a beneficial allele—that is an allele that enhances its evolutionary fitness—or, conversely, it can add an allele that is detrimental to the population's fitness.

The Effect of Catastrophes on Small Populations

Although no population is immune to genetic drift, small populations are far more vulnerable because their gene pools have fewer alleles. Sharp drops in population size—say, because of war, or disease, such as the flu pandemic of 1918—can also change allele frequencies substantially, even to the extent of weeding out an allele entirely. Biologists call this phenomenon a "crash." When a population crashes, the alleles in the surviving sample may not be representative of the gene pool that was present before the calamity struck.

For example, consider a population of worms, half of whom are wiped out by a flood. Now suppose that a particular allele is carried by 25 percent of the worms. In an original population of 100,000 worms, 50,000 worms will survive and we can expect that a little more than 12,000 worms will have the allele. But if the population of worms were very low—say, 12 or so—the laws of probability predict that only 1.5 of the survivors would have the allele. So, you can see how, in these circumstances, the allele can be eliminated from the population entirely.

What Does It Mean?

A **population bottleneck** refers to a dramatic change of environment, such as disease or famine, which reduces a population's size and genetic diversity.

Running Out of Steam: Bottlenecks and the Founder Effect

The flood in our example caused what biologists call a *population bottleneck*. This term refers to the effect of reducing the genetic diversity in a small population; even if it manages to grow back to its original size, it will have an impoverished genetic inheritance. That's because only a fraction of the alleles in the original gene pool will remain. If the survivors succeed in starting a new population, they have fewer alleles to contribute to it. This is known as the *founder effect*. A founder effect can be seen in the northern elephant seal population, which was hunted nearly to extinction in the nineteenth century. With a ban on hunting, their numbers have rebounded, but the population bottleneck, caused by the hunting, has left them with fewer alleles to pass along to the next generation.

Why should it matter whether there are fewer alleles in a given population? A partial answer is provided by another example—the Amish of Pennsylvania. All

What Does It Mean?

The **founder effect** refers to the diminished genetic diversity of a species that tries to re-establish itself after a natural calamity has decimated its numbers. This effect is usually seen in small populations where the loss of genetic characteristics can be more pronounced than in larger ones.

the Amish are descendants of about 200 people who established their community in the early 1700s after migrating from Europe. One of the founders had a rare allele that causes a type of dwarfism. Now 1 in 14 Amish have this same dwarfism, compared to 1 in 1,000 in the general population. Populations with fewer alleles are more subject to disease or environmental pressures because they lack the diversity to cushion the blow. Panthers in Florida, for instance, are all virtual clones of one another—they are almost all of the same genetic stock. That means that a bacterial or viral infection, for instance, that can kill off one animal is capable of killing off the entire population. On the other hand, the human population—six billion and counting—has so much genetic diversity that a certain percentage of humans are likely to have an immunity to virtually any outbreak of disease. (Some people, for instance, have a natural immunity to HIV, which causes AIDS.)

How important is genetic drift in the overall evolutionary scheme of things? Fisher thought that populations were sufficiently large that drift played a negligible role. The American geneticist Sewall Wright disagreed, contending that the focus should be on smaller subpopulations where drift could make a significant difference.

Both natural selection and genetic drift tend to decrease genetic variation. The former selects certain genetic variants, gradually culling them out of a population, while the latter can remove alleles from a population by chance. Except in rare instances, new alleles enter the gene pool as a single copy. Traveling solo means that most new alleles are lost almost immediately as they join the gene pool due to drift or natural selection; only a small percent ever reach a high frequency in a population.

The effects of natural selection and drift complement each other. Genetic drift will become intensified as selection pressures increase. For example, when a change in the environment—the hunting of northern elephant seals or the drowning of a population of worms, to use two examples that we cited earlier—occurs, natural selection goes to work, culling the individuals who lack the fitness to adapt to the new circumstances. That reduces the effective population size and means a diminishment of the number of individuals available to contribute alleles to the next generation. And, as we noted, when you have a smaller population, you usually have fewer alleles available. That is, genetic diversity is lessened. When there's too little genetic variety, the population is more at risk for disease and other environmental disasters.

What Keeps Evolution in Motion?

So if natural selection and genetic drift were the only mechanisms of evolution, populations would eventually become homogeneous—they would have very similar genetic makeup, like the Florida panthers—and then you could imagine a situation in which further evolution would become impossible. But further evolution isn't impossible, as we know. It is going on all the time. So, that means there must be other mechanisms that increase variation depleted by selection and drift. How variation is introduced into a population to maintain genetic diversity (and, incidentally, keep evolution in motion) is a subject that we'll be taking a look at in our next chapter.

The Least You Need to Know

➤ Natural selection is only one mechanism of evolution.

➤ While natural selection is "blind"—it has no overriding goal apart from enhancing organisms' adaptive ability—artificial selection, guided by humans, has definite objectives, such as producing a faster racehorse, generating a greater yield from a crop, or developing a cow that will give more milk.

➤ Many secondary sexual characteristics that have seemed favored by natural selection do not seem to confer a survival benefit and may even be detrimental to survival.

➤ The danger posed by secondary sexual characteristics—they may make it easier for predators to find their prey—may be outweighed by their advantage for reproductive success.

➤ Genetic drift, in which alleles are added or subtracted at random from a population, usually results in a loss of genetic diversity because new alleles have very little chance of taking hold (or becoming fixed) in a population.

➤ Both natural selection and genetic drift remove alleles from a population and act together to reduce population size (and, thus, its genetic diversity).

Chapter 15

Mutants, Shuffled Genes, and Mateless Maggots

In This Chapter

➤ Mechanisms that increase genetic diversity

➤ Mutation: mistakes that matter

➤ Why bad alleles survive

➤ Gene flow: the impact of migration

➤ Recombination: sharing genetic material

In our last chapter, we examined the ways in which genes are plucked out of the gene pool by means of natural selection and genetic drift. Both of these phenomena play a vital role in evolution—one by selecting favorable traits and rejecting the ill-adaptive traits, and the other by randomly changing the genetic mix. Even though new genes, whether beneficial, harmful, or neutral, occasionally enter a gene pool, they come alone and so are vulnerable to being eliminated before they have a chance to enjoy their new home.

The only reason that new genes do eventually take hold in a population—become fixed, as biologists like to say—is because they keep recurring through mutations. It's like playing the same numbers in a lottery long enough, in hope that, sooner or later, the odds will turn in your favor. As we'll see, mutations are only one of the methods nature employs to replenish the gene pool with new genes. Genes are added as well through gene flow—the migration of other reproductively compatible species—into a population and through recombination, a random mixing of genetic material between strands of chromosomes.

Although scientists now understand many of the mechanisms that propel evolutionary change, they have yet to work out how exactly new species are created on a genetic level. (Not that they don't have many theories and speculations—scientists almost always do.) Surprisingly, there's even considerable debate on how important the creation of new species—called speciation—actually is in the overall scheme of evolutionary change. Nonetheless, scientists have acquired a very good understanding of what needs to happen to produce a new species. In fact, they have observed speciation taking place, at least in certain species of plants.

The key to speciation is reproductive isolation. Reproductive isolation is what happens when groups of a certain population are no longer capable of interbreeding. Reproductive isolation takes place in two ways: Groups in a population become geographically isolated from each other and develop independently, or they become reproductively incompatible because of genetic mutations or other changes, even though they still remain in proximity.

Putting Genes Back in the Pool

In our last chapter we talked about how genetic drift and natural selection actually deplete genetic variety in the gene pool. So what keeps the ball in motion? For evolution to take place at all a generous supply of genetic variants needs to exist in the population. So we would expect there to be other mechanisms that nature employs to replenish the gene pool, some of which we will consider in the following section.

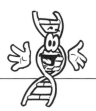

A Natural Selection

DNA, deoxyribonucleic acid, is made up of a large number of chemical compounds called nucleotides, consisting of a sugar, a phosphate group, and one of four different nitrogen-bearing compounds called bases. These bases—adenine, guanine, thymine, and cytosine—are often designated by letters: A, G, T, or C.

Mutation

The cellular machinery that copies DNA sometimes makes mistakes, which alters the sequence (the chemical composition) of a gene. This is called a mutation. There are many kinds of mutations:

➤ A mutation in which one "letter" of the genetic code is changed to another, a phenomenon known as point mutation

➤ A mutation in which lengths of DNA are deleted or inserted in a gene

➤ A mutation in which genes or parts of genes can become inverted or duplicated

Most mutations are thought to be neutral with regard to fitness. That is, they don't affect reproductive success. Generally, most random changes caused by mutations will not produce a fitness benefit. In fact, only a very small percentage of mutations are beneficial.

Harmful mutations are more common, especially those that have any phenotypic effect. These types of mutations alter appearance or the physical structure of the organism or its offspring. Cystic fibrosis, for instance, is a hereditary disease that has a phenotypic effect because it produces an abnormal secretion of thick mucus that clogs the pancreas and lungs, leading to respiratory and digestive problems that may result in death.

At this point, let's stop for a moment to consider what DNA actually does in the body and why any interference with its function or composition has the potential of doing considerable damage:

➤ DNA is carried in genes on chromosomes in every cell of an organism's body. The DNA contains instructions that are ultimately responsible for an organism's traits and behavior. Genes act separately and in combination to produce their effects.

➤ The DNA contains instructions for the production of proteins. Proteins are organic compounds that are essential to the functioning of living organisms.

➤ Proteins are made up of smaller molecules called amino acids. The sequence of amino acids—its chemical composition—is determined by the sequence of the nucleotide bases in DNA—adenine, guanine, thymine, and cytosine.

That's why mutations can be so dangerous. Change a C to an A in the DNA, or lengthen a strand of DNA or shorten it, or substitute one amino acid for another, and you can trigger a cascade of abnormal biochemical reactions. In this manner, a mutation can throw a very delicate system out of balance. Mutations tamper with the genetic instruction manual. Instead of telling a protein or enzyme to stop doing something, the alteration has produced an instruction that now says it's perfectly okay to keep going! Green lights turn to red, and red lights stay on too long. The result can be havoc. Some mutations, for example, are implicated in several serious diseases, ranging from depression to cancer.

Why Do Bad Alleles Endure?

Mutations, especially harmful ones, inevitably occur in various populations. All human populations have deleterious mutant genes. Many of these mutations, though, are not so harmful that they have been eliminated by natural selection. If mutations were so harmful that they imperiled the survival of a population and jeopardized its reproductive efficiency, they wouldn't stick around for very long. If the population doesn't survive or reproduce, obviously the mutation would vanish altogether. Natural selection that weeds out harmful alleles is called negative selection. Some harmful mutations endure because they know how to behave themselves under most (but certainly not all) circumstances.

Deleterious alleles are mutants that remain in the population in spite of their bad effects. Mendel called these alleles recessive; when paired with normal alleles, the bad effects don't show up, so often the person carrying them doesn't realize he has them. You can be a carrier of a disease without ever having to suffer from the effects of the disease. Recall our sickle cell example. Some scientists also speculate that if genes for depression are discovered, they may turn out to have some advantage as well, but perhaps only when paired with a dominant allele. That would suggest that many more people in the population have genes for depression than actually get the illness. Natural selection doesn't act against these recessive alleles for the simple reason that, as far as selection is concerned, they are "invisible," masked by a dominant allele. As a result, these recessive alleles—the deleterious kind—slip through the cracks of evolution. Unless one carrier mates with another carrier, the allele may simply continue to be passed on generation after generation, undetected.

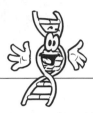

A Natural Selection

Although genes for depression have yet to be found, substantial evidence exists for calling it a genetic illness because depression and manic depression tend to turn up in several generations of the same family. Given the persistence of depression in the human population, scientists and philosophers speculate that the genes responsible may confer some kind of benefit, which is still unknown.

Negative Selection: Eliminating Mutant Genes from a Population

Some mutant genes, on the other hand, are eliminated by natural selection. The elimination of harmful mutants is referred to as *genetic death*. Genetic death can come about in various ways—some quite ruthless. The death of a child from a serious hereditary disease could be considered a genetic death in that it eliminates the mutation before it can be passed on. But genetic death can also take another form, such as the failure of the parents of the child who has died to have another child. These are both considered genetic deaths because they subtract genes from the population or fail to add them.

What Does It Mean?

Genetic death refers to the elimination of a bad gene variant or allele from the population; it can occur when a child is unable to survive to reproductive age because of a hereditary disease (eliminating the allele that causes the disease) or when parents with the bad alleles fail to reproduce.

When mutations occur at a higher frequency, more of them are likely to be harmful. This makes sense if you think about it. The more mutations that occur, the more mutations of all kinds will be produced—both good and bad. The sheer number of bad mutations will rise. And many of these mutations will indeed be harmful enough to affect survival and fitness, so they will be selected out of the population at a greater rate. This is another way of saying there will be more genetic deaths. Think of what happened at Hiroshima or Chernobyl, where populations were exposed to lethal dosages of radiation, one as a result of the atomic bomb and the other as a result of a meltdown at a nuclear power plant. Many people in each location who survived the initial exposure went on to develop genetic mutations; some became unable to have children, and others miscarried or bore children with congenital deformities, all instances of genetic death.

While harmful genetic mutations are being eliminated from a population by natural selection, other harmful mutations are being accumulated by chance. (As we noted earlier, mutations occur usually in reproduction, although some can arise from environmental causes, such as radiation or exposure to toxic chemicals.) At a certain point, the numbers of the mutant genes created become equal to the numbers of mutant genes being eliminated. The population is then said to be in the state of genetic equilibrium.

Mutations don't necessarily have much effect on populations over short time spans. Ordinarily, it takes a long time for a mutation to take hold (or become fixed) in a population. So, it's possible that populations are never at equilibrium. However, a change in environment can cause neutral alleles that previously had no effect one way or another to become beneficial and enhance adaptation. It's as if you were given stock in a widget company that was barely worth the paper it was printed on. Then a few years later, widgets are suddenly in high demand, your previously worthless stock soars, and you congratulate yourself on your wisdom in keeping hold of it.

We can say the following about mutations:

➤ Mutations may be neutral or produce good or bad alleles. Mutations usually occur in reproduction.

➤ Some bad alleles survive because they are recessive and are thus "invisible" to natural selection since they are ordinarily paired with a dominant allele.

➤ Some bad alleles may confer an adaptive advantage to the person who has them although usually only if they are paired with a dominant allele, such as the protective advantage against malaria that occurs when a sickle-cell allele is paired with a normal one.

➤ Some mutations are eliminated by natural selection, as in the case of certain hereditary diseases. This phenomenon is called genetic death.

➤ When mutations occur at a higher frequency in the population they are more likely to be harmful than good or neutral.

Good Alleles

Although they may be relatively rare, new mutants do occur. But, like deleterious alleles, they may get lost—they fail to take or become fixed in the population. On the other hand, if mutations occur often enough in a given population, the chances increase that the beneficial mutation will show up again. After several failures, it may finally stick.

Take the case of a species of mosquito, *Culex pipiens*. Because of a chance mutation, a gene that the mosquito had involved in breaking down organophosphates—common insecticide ingredients—became duplicated, essentially giving the mosquito more bang for its buck. The duplicated gene made the mosquito far more resistant to DDT and other insect repellants. It didn't take long before the progeny of *Culex pipiens*, now happily endowed with this mutation, to overrun much of the nonmutated mosquito population worldwide because of its heightened resistance.

Now you may not think that this mutation was necessarily beneficial when you have a cookout, but from the mosquito's point of view, it was equivalent to winning the lottery. Natural selection that increases the frequency of beneficial alleles is called *positive selection*, or positive Darwinian selection.

All things being equal, though, beneficial mutants occur infrequently. Even most moderately beneficial alleles are lost because of genetic drift when they appear. As a consequence, just about all the fitness differences in any population will be due to new deleterious mutant genes and the deleterious recessive alleles. In these circumstances, Natural selection will be limited to a role of screening out unfit variants.

What Does It Mean?

Positive selection refers to the increase of the frequency of beneficial alleles in a population because of natural selection. It is a phenomenon that occurs rarely.

Gene Flow: A Little Help from One's Friends

Evolutionary change can occasionally occur when new organisms of the same or a closely related species enter a population because of migration. If they mate within the population, they can add new alleles to the local gene pool. This is called gene flow. In some closely related species, fertile plant hybrids can result from matings between the species. When the species are more distantly related, gene flow occurs infrequently.

Recombination: The Great Gene Shuffle

Recombination is a form of gene shuffling that produces new genetic combinations that can express themselves in new forms of visible traits in the offspring. First, we need to provide a little background in reproductive biology. Each chromosome in our sperm or egg cells is a mixture of genes from our mother and our father. (Chromosomes, remember, are microscopic, highly complex structures in the cells of most organisms that contain DNA.) Most organisms have linear chromosomes (with the notable exception of bacteria), with their genes lying at specific location (loci) along the strands. Every cell of most sexually reproducing organisms contains two of each chromosome type—one from each parent. The sex cells are distinctive, however, in that they contain only half the normal number of chromosomes.

What Does It Mean?

Meiosis is a process of cell division during which the cell's genetic information, contained in the chromosomes, is mixed and divided into sex cells. These sex cells, called gametes, have half the number of an organism's chromosomes. In fertilization, the sex cells of both parents combine to form offspring with the full number of chromosomes.

In a process of cell division called *meiosis,* the cell's genetic information, contained in chromosomes, is mixed and divided into sex cells called *gametes.* Pairs of chromosomes may exchange their genetic information in a process called recombination or crossovers, which occur more or less at random along the lengths of the chromosomes. The random sorting of chromosomes during meiosis ensures that each new sex cell will provide the offspring with a unique genetic inheritance, a mixture of the parents' genetic information. During fertilization, the sex cells from one parent will combine to form offspring with the full number of chromosomes—half from the mother and half from the father. The cell resulting from the union of two gametes (sex cells) is called a *zygote.* The zygote will usually undergo a series of cell divisions until it develops into a complete organism. In the offspring produced by the gametes, the crossovers show up as new combinations of visible traits.

Evolution is a change in the gene pool of a population over time that can come about because of several factors. Three mechanisms add new alleles to the gene pool: mutation, recombination, and gene flow. Two mechanisms—natural selection and genetic drift—remove alleles from the gene pool. Selection removes deleterious alleles from the gene pool, while drift removes alleles at random from the gene pool, regardless of whether they're favorable. The amount of genetic variation—whether there is more or less diversity in a population—is based on the balance between the actions of these mechanisms. In other words, if natural selection and genetic drift are more predominant, there is likely to be less genetic variation in a population;

What Does It Mean?

A **gamete** is the sexual reproductive cell that fuses with another sexual cell in the process of fertilization. A **zygote** is the cell resulting from the fusion of a male and female sex cell. The zygote will need to undergo many cell divisions before it develops into an organism.

171

if mutation, recombination, and gene flow are more predominant, you would expect more genetic variation.

The following is a list of some of the most important mechanisms for genetic change:

➤ Natural selection

➤ Artificial selection

➤ Sexual selection

   The good gene model

   The runaway sexual selection model

➤ Genetic drift

➤ Mutation

➤ Gene flow

➤ Recombination

Where Do New Species Come From?

Natural selection, genetic drift, gene flow, and recombination continuously influence and change the characteristics of a population. Yet, in many instances, the influence is relatively slight and seldom is sufficient to create an entirely new species. But, of course, the creation of new species is something that has preoccupied scientists for centuries, long before Darwin. (Although Darwin's major work was titled *The Origin of Species,* he avoided addressing the issue directly.) It's one thing to understand how these various evolutionary mechanisms operate. But how you get to the next step—the development of a new species from an older species—is not so easily comprehended.

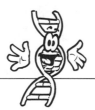

A Natural Selection

The equid family consists of horses, zebras, and asses, all of which diverged from a common ancestor about four million years ago; they all interbreed with one another, but the offspring are almost invariably fertile.

The bottom line in the creation of a new species is when, for whatever reason, members of a population cease to interbreed. When populations can no longer mate, they are said to be reproductively isolated from one another. That means that two reproductively isolated populations cannot randomly exchange genetic material with each other. Reproductive isolation sets in motion a process in which the incompatible groups diverge and evolve independently of one another. This process is called speciation.

All things being equal, interbreeding in a population will continue as long as nothing comes along to stop it. Any factor that hinders interbreeding is called an isolating mechanism. As we noted earlier, geographic barriers—mountains, oceans, deserts, and so on—will act as isolating mechanisms. When populations are fragmented in this way, the process may lead to the formation of entirely new species. Once separated from one another by such an isolating mechanism, the gene pools of the two populations change to such an extent that the populations wouldn't be able to interbreed even if they were reunited. This process of creation by isolation is called allopatric speciation. Darwin observed several examples of allopatric speciation when he visited the Galapagos. Even though the distances were not very great between islands, the finches, lizards, and tortoises had all developed into separate species.

A second kind of speciation occurs, although less frequently, when mutations or subtle changes in behavior prevent individuals living close together from reproducing, a phenomenon known as sympatric speciation. Sympatric speciation occurs when two distinct subgroups of a population are no longer capable of exchanging genetic material and evolve into two or more distinct species. This kind of speciation can be seen in insects that live on a single host plant. If you divided the insects into two groups and transferred one group to another host plant, the second group eventually would no longer be able to mate with the group inhabiting the original host plant. That is, the two subpopulations would diverge and speciate. Agricultural records dating back to the 1860s showed that a strain of apple maggot fly—*Rhagolettis pomenella*—had previously infested only hawthorn fruit. Now the maggot flies that infest apples are unable to mate with the descendents of their forbears who still crave the hawthorn fruit.

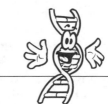

A Natural Selection

Maggots are any worm-like larva of a fly, especially those laying their eggs in decaying materials. The rotting material provides insulation for the hatching of eggs and food for the newly born maggots.

The basics of speciation are as follows:

➤ New species arise when members of a population can no longer interbreed successfully and exchange genetic material.

➤ Anything that prevents interbreeding from taking place is called an isolation mechanism.

➤ Species can be isolated in two ways—allopatric and sympatric:

　➤ Allopatric speciation is the creation of a new species when different groups of a population become physically isolated from one another, causing them to diverge and evolve separately.

　➤ Sympatric speciation, a more uncommon mode of speciation, is the creation of a new species when groups in a population living close together are unable to interbreed because of a mutation or subtle behavioral changes.

As much as biologists have discovered about genetics, they still are in the dark about the precise genetic mechanisms of speciation. Some are convinced that speciation gradually occurs through a series of small changes in each subdivision of a population until, one day, the groups are incapable of mating. One possible scenario is based on the founder effect (which we discussed in the last chapter), in which the remnants of an environmental disaster try to reestablish a community, often with a diminished genetic capacity. The descendants of these survivors might undergo relatively rapid speciation—a virtual genetic revolution—through the transformation of a few key genes that impose reproductive isolation on them. Because populations of organisms are very complicated, it's likely that speciation can occur in many ways. More than 150 years after the publication of *Origin* (1859), the question of how species originate is still largely a mystery.

Even though scientists may not be sure how new species arise, speciation is an established fact, observable in insects and plants. Many scientists think that speciation is key to understanding evolution. How important speciation is, however, remains a source of dispute. Some scientists, including paleontologists (who study fossils), contend that large-scale evolutionary change—macroevolutionary change, as it's called—cannot occur without speciation. A number of geneticists, on the other hand, do not believe that speciation is essential for major evolutionary change to occur.

The Least You Need to Know

➤ Genetic mutations may be harmful, beneficial, or neutral.

➤ Many harmful alleles, in combination with normal alleles, may confer a benefit and thus are conserved by natural selection.

➤ Gene flow is the migration of new genes into a population, brought in by species that are capable of breeding with the established population.

➤ Recombination is a shuffling of genes through a process of sharing genetic information between chromosomes (cellular structures containing DNA); this recombined information is then passed along to offspring that will express the new traits.

➤ Scientists have established that new species are created when members of a population are no longer capable of interbreeding with other members of a population because of reproductive isolation.

➤ Reproductive isolation can occur in two basic ways: allopatric, in which a group becomes geographically isolated from its original population, or sympatric, in which one group can no longer interbreed with other members of the population because of a mutation or some subtle behavioral change.

Part 4

Putting It Altogether: The Modern Synthesis

While geneticists were struggling to integrate genetics with Darwinian theory to understand evolution on a larger scale than previously thought possible, other disciplines were taking very different approaches in their investigations. Evolution was—and is—such a wide open field that everyone wanted to get a piece of the action: botanists, geologists, paleontologists, anthropologists, and even theologians.

For instance, it was a German astronomer and meteorologist turned geologist named Alfred Wegener who came up with the theory of continental drift, with far-reaching implications for evolution. Wegener theorized that the six continents that exist today had once constituted a single "supercontinent" that was split apart hundreds of millions of years ago by geological upheaval. Wegener's ideas initially provoked outrage and scorn because they flew in the face of conventional wisdom. But then, of course, controversy is nothing new to evolution.

In Part 4, we see how controversy has also overshadowed the efforts of anthropologists and paleontologists to trace the origins of humankind in Africa. A succession of startling discoveries (that continue to this day), far from clearing up the mystery of how humans evolved, seemed only to raise more questions than they answered.

Chapter 16

The Big Picture: Macroevolution

In This Chapter

➤ Microevolution: change within a species

➤ Macroevolution: change above the species level

➤ Why evolution isn't progress

➤ Proving common ancestry

➤ The pace and timing of evolutionary change

As we've seen, many naturalists and philosophers before Darwin cherished the idea that evolution represented progress, in which simpler, more primitive forms of life evolved over time into more complex organisms, with humans representing the culmination of the process. This idea was exemplified in the chain of being. But evolution doesn't work in a straight line. On the contrary, evolution resembles a branching tree or a bush. Perched on the tips of each branch are the several million species that are living today. The contrary view that evolution is progressive, or that it has a direction or an objective, is called orthogenetic.

Without the theory of evolution, the strategies that many species pursue would be difficult to comprehend. Evolutionary change couldn't be understood, either, without knowledge of how the Earth was shaped, how oceans were carved out, and how land masses split off from one another. Geology had another role to play as well—setting back the clock. The Earth, it turns out, is more than four and a half billion years old (you recall how Bishop Ussher estimated that only 5,000 years had passed since creation). Then, about three and a half billion years ago, a humble single-cell organism showed up—the first life on Earth.

It wasn't until relatively recently, however, with the accumulation of the fossil record and the development of molecular biology (enabling scientists to study genetic variation and similarities in a variety of organisms) that scientists could begin to appreciate just how the evolutionary tree came to be. So far, we've been examining how species evolve and how new species are created. But what happens above the species level? Some framework was needed that scientists could use to describe what went on among the different branches of the tree. That led to the development of what is known as macroevolutionary theory.

Followers of macroevolutionary theory believe, with substantial evidence to bolster their contention, that the same evolutionary processes that take place on the microevolutionary level—that is, within a species—also takes place *between* the species as well. Using fossil and genetic evidence, scientists were able to demonstrate conclusively how species shared common ancestry and establish why certain organisms have organs or behaviors that, serving little or no purpose at present, are left over from ancestral species.

Macroevolution theory also gave scientists a foundation to tackle some tough and still unsolved questions. For instance, why did the fossil record show that many species remained relatively stable for long periods of time and then suddenly underwent a rapid period of evolutionary change that didn't seem to leave any transitional organisms in its wake? Why did evolution accelerate at one time and not at another? Could a change in a species affect a larger population composed of many species?

What Does It Mean?

A clade is a group such as a species, whose members all share a common ancestor. **Cladogenesis** is the splitting of a lineage into two species because members of a population sharing a common ancestor are no longer reproductively compatible.

Macro- and Microevolution: What's the Difference?

In science, the prefix *macro-* simply means "big," while the prefix *micro-* just means "small." (Both are derived from Greek words.) In evolutionary biology, macroevolution is a term used to refer to any evolutionary change at or above the level of species. Macroevolution takes place, for instance, when speciation occurs—that is, when populations are divided into distinct species that are no longer able to reproduce with each other. This results in yet another branch on the tree of evolution; the term biologists use to describe the split is *cladogenesis,* from the Greek meaning "the origin of a branch." Similarly, any changes that occur at higher levels, such as the evolution of new families, phyla, or genera, are also considered macroevolutionary.

The classification system that scientists use to categorize organic life is based on seven categories, listed from the largest to the smallest. Every known organism occupies a specific place in each group.

➤ **Kingdom.** There are five kingdoms: animals, plants, fungi, protist (including algae, ciliates, sporozoans, and flagellates), and prokaryotae (bacteria and some algae).

➤ **Phylum or division.** (Phyla are used to designate animals, but the term division is applied to plants, fungi, and protist.) The animal kingdom is divided into 20 or more phyla. All animals with a backbone, for instance, belong to the phylum Chordata. Plants are divided into two phyla. Plants with flowers, for instance, belong to the division Tracheophyta.

➤ **Class.** Members of a class have more characteristics in common than in either a kingdom or a phylum. Mammals, reptiles, and birds all belong to a single phylum, for example, but fall into a different class. Apes, monkeys, and mice are all in the class Mammalia.

➤ **Order.** Members of an order share more characteristics than members of a class. Dogs, moles, raccoons, and shrews, for instance, are all mammals, but only dogs and raccoons belong to the class Carnivora because they are flesh-eaters. Shrews and moles are classified as Insectivora.

➤ **Family.** Members of a family have more in common than organisms in a class. Wolves and cats belong to the class Carnivora but are placed in different families: The former belongs to the family Candae, and the latter belongs to the family Felidae.

➤ **Genus.** This category includes very similar groups, but members of the groups usually cannot breed together.

➤ **Species.** This is the basic unit of biological classification. Members of a species have many common characteristics and are capable of breeding with each other.

This classification system—partly based on the one that Carl Linnaeus devised in the 1700s—not only helps scientists distinguish the differences and similarities of existing organisms, but it also provides them with a road map that allows them to see how species share traits that they inherited from a common ancestor. The search for a common ancestor is one of the principal aims of macroevolutionary theory. However, when you break down the larger groups—phyla, families, classes, and so on—into subgroups, you'll find unique traits peculiar only to that group.

Similarities are what bind groups together. Their differences provide important clues on how they are subdivided today and when the division occurred that separated one group from another. For example, plants can be divided into two broad categories: nonvascular (mosses, for example) and vascular (having a circulatory system). Vascular plants, in turn, can be divided into seedless (ferns, for example) and seeded.

A Natural Selection

Vascular plants make up a phyla characterized by internal conducting tissue for the movement of water, minerals, and food.

Vascular seeded plants can be divided once again—into gymnosperms (pines, for example) and flowering plants (angiosperms). Angiosperms can be broken down into monocots and dicots. And it doesn't stop there: Monocots and dicots have several distinctive characteristics that distinguish them from other plants.

It turns out that traits are not mixed and matched willy-nilly in groups of organisms. For example, flowers are seen only in plants that possess certain characteristics. Flowers are found in angiosperms, for instance, but not in gymnosperms. This is the kind of pattern that scientists expect to see in looking for evidence of common descent.

Evolution Above the Species Level

So we can define macroevolution as evolution that happens above the species level. When scientists look at why some species expanded and others dwindled over time, or try to understand why insects developed wings, or puzzle over how evolution was affected by geological change, they are asking questions that belong in the realm of macroevolutionary theory. In macroevolution, scientists seek insights into the processes that have shaped the major evolutionary patterns of present-day and ancient organisms. Macroevolutionary theory also helps explain how organisms are modified over time by the cumulative effect of natural selection. The theory studies the way in which genetic traits are distributed among different populations to determine how lineages split off from one another and how new or novel traits are produced by mutation. As a multidisciplinary field, macroevolution relies on contributions from biologists, geneticists, paleontologists, molecular scientists, geologists, and even astrophysicists.

Evolution Below the Species Level

Microevolution, by contrast, refers to any evolutionary change below the level of species. Microevolution considers changes in the frequency of alleles of a population or a species and how those changes affect the form or appearance (the phenotype) of organisms that make up that population or species. Darwinism forms the basis for microevolutionary theory. That's where natural selection figures into the equation. Natural selection, as we said, optimizes the existing genetic variation in a population to enhance its reproductive success. In that sense, microevolutionary theory offers scientists a reliable framework to interpret the frequency and relative importance of a variety of traits. In microevolutionary theory, scientists try to weigh the trade-off between natural selection and sexual selection. What sacrifices do species make to attract a mate?

Micro- and Macroevolution: Defining the Relationship

Stated another way, macroevolution traces genetic change among different species, whereas microevolution looks at genetic change within a single species. Micro-evolution can be studied directly. Think of Darwin's observations on the Galapagos or the speciation that occurs when one group of insects is separated from another and placed on a different host plant. That's microevolution in action. Macroevolution cannot be directly observed. To study macroevolution, you need to be a detective and examine all the evidence and data that you can possibly gather—and then make your deductions accordingly.

The Modern Synthesis

Macroevolution represents what's known as the "modern synthesis" of neo-Darwinism, which developed in the period from 1930 to 1950. (*Neo* means "new." Neo-Darwinists are scientists who accept Darwin's theories as their foundation and work from there.) Macroevolutionary theory is considered a modern synthesis because it reconciled Darwin's idea of evolution by natural selection and the discoveries of modern genetics. Modern genetics traces its development to the ground-breaking work done by Watson and Crick on the structure of DNA beginning in 1953.

The terms *macroevolution* and *microevolution* were first coined in 1927 by Russian ento-mologist (insect researcher) Iurii Filipchenko, who was trying to reconcile evolution and Mendalian genetics. (He was not, however, a Darwinian.) His student, geneticist and zoologist Theodosius Dobzhansky, introduced the terms to the United States in his 1937 book *Genetics and the Origin of Species*, which he began by saying that "we are compelled at the present level of knowledge reluctantly to put a sign of equality be-tween the mechanisms of macro- and microevolution" His reluctance stems from his realization that he was proposing a theory at odds with that of his beloved teacher.

Who Are They?

Theodosius Dobzhansky (1900–1975) was a Russian-born geneticist and zoologist who made major contributions to the field of population genetics, based on his studies of the fruit fly. His findings served as the basis for his theories as to how humans may have evolved through adaptation.

As a concept, macroevolution fell into obscurity throughout the 1940s. Most neo-Darwinian writers preferred to consider evolution in terms of changes in allele frequencies without getting involved in evolutionary processes beyond the level of species. In recent years, though, macroevolution has regained popularity, especially in the United States through the writings of the celebrated Harvard paleontologist Stephen Jay Gould (more about him in Chapter 17, "The Role of Extinction in Evolution").

Synthesists, who embrace macroevolutionary theory, contend that the same processes that cause microevolution (changes of the frequencies of alleles within a species) can be applied to macroevolution (changes between species). They basically dared their critics (and there were many) to prove the opposite: that the processes weren't connected, that what went on within species operated under one set of rules, and that what went on between species was governed by another set. Was there some mechanism to prevent microevolution from causing macroevolution? But no compelling scientific evidence has yet turned up to suggest that such a mechanism exists. To put it another way, the rules for microevolution and macroevolution are the same; they just apply on different scales.

The modern synthesis still isn't accepted by all scientists, however. Some evolutionists who have never signed on to Darwin's theories maintain that the processes that cause the development of new traits within a species are different from the processes that cause speciation. Darwin, on the contrary, thought that the very processes that created new traits were also involved in creating new species. If enough new traits developed in certain subgroups of a population, they would no longer be able to reproduce with members of the original population and diverge into a new species. Evolution is not progress. Still other scientists disagree on different grounds, insisting that some as yet unknown mechanism is responsible for macroevolution.

The Dynamics of Macroevolution

Given the scope that macroevolutionary theory has set for itself, it should scarcely come as a surprise to learn that there are several different conceptions of how it works. Punctuated equilibrium theory, for instance, proposes that once species have originated and adapted to their environmental circumstances, they tend to stay pretty much as they are for the rest of their existence before undergoing rapid change. In contrast, *phyletic* gradualism says that species continue to adapt to new challenges over the course of their history. Then there are theories—one called species selection and another called species sorting—that consider how long species can stick around before wearing out their welcome. According to these theories, as a result of macroevolutionary processes, certain species are either more or less likely to exist for very long periods before becoming extinct.

What Does It Mean?

Phylectic change refers to gradual evolutionary change; it is also known as vertical change.

Looking for Common Ancestors

Advocates of macroevolution are basically in agreement that species are descended from common ancestors. As time went on, different lineages of organisms underwent modification into new species as they descended from a common ancestor. This modification took place as species adapted to their environments. No living organisms today are our ancestors. No species is more "modern" than any other. All species can claim their own unique evolutionary history. Both humans and apes share a common ancestor, and both are also fully developed modern species. It's true that the ancestor we evolved from was an ape-like creature, but that creature vanished long ago and bears little more resemblance to present-day apes than it does to humans.

How can you determine that several existing species come from a common ancestor? The best evidence comes from five major sources:

➤ Genes

➤ Structural (morphological) similarities

➤ Fossils

➤ Geographical proximity of fossils

➤ Similarities in embryonic development

A Natural Selection

Apes originated in Africa about 20–22 million years ago; about 15 million years ago, some apes migrated to Europe over a land bridge that no longer exists.

Genes contain the history of an organism. You've probably heard of "genetic fingerprinting." Because each person's genotype (underlying genetic makeup) is unique, his DNA can act as a "signature." That signature is embedded in every cell of your body, including flakes of skin and follicles of hair. Recently, a number of men convicted of rape have been set free because they were exonerated by DNA evidence. In the same way, DNA can lead scientists in a quest to discover genetic relationships that would otherwise have eluded detection. At the same time, using the same kind of DNA techniques, scientists can travel back in time hundreds of thousands or even millions of years to identify ancestors of existing species that may have long since vanished.

We commonly say that a scientific theory is valid if we can use it to make testable predictions and then see whether the same results hold up under the same circumstances each time the test is performed. For example, if two organisms share a similar anatomy, you could predict that their gene sequences would have more in common than two organisms that bear little or no resemblance to each other. (In that case, scientists say that the organisms were *morphologically distinct*.) With gene-sequencing techniques now being used—which rely not only on microscopes but also on vast arrays of computers for number crunching—geneticists have discovered that, as they had suspected, the correspondences between body structure and genetic makeup predicted by evolution are usually very high. That suggests to them that common ancestry must be involved on two counts: The structures all seem to have developed along similar paths, and the organisms all shared the same genetic information to some degree.

To determine whether species are closely related—that is, whether they come from a common ancestor—biologists look for a similarity in structure (morphology) and a similarity in its genetic sequence. Gratifyingly, morphological similarity does provide a fairly reliable basis for predicting that the gene sequences will also be very much alike. That's not to say that the gene sequences are exactly the same—some differences do exist, but they are usually not significant.

We can say that groups of related organisms are "variations on a theme." Take the bone structure of vertebrates. All vertebrates are built on the same model. As we noted earlier, the bones of the human hand grow out of the same tissue as the bones of a bat's wing or a whale's flipper. In addition, they share many identifying features, such as muscle insertion points and ridges. Admittedly, they are scaled differently, and they serve entirely different purposes. Nonetheless, evolutionary biologists have concluded from such evidence that all mammals are modified descendants of a common ancestor that had the same set of bones. In addition to genetic and morphological similarity, biologists look for geographic proximity—often based on the fossil record—as an indication that organisms are closely related.

Although the criterion for deciding whether individuals belong to the same species is clear, making the distinction in practice isn't always so easy. For one thing, it may not be known for certain whether individuals living in different locations belong to the same species because it is not known whether they can naturally interbreed. If they cannot interbreed, they cannot belong to the same species, by definition. The other reason that scientists sometimes have trouble determining whether a close relationship exists is rooted in the nature of evolution itself. Evolution is a gradual process in which new species emerge from established ones, eventually becoming incapable of interbreeding with one another. Because the process is gradual, though, it's often difficult to say where the cut-off point is when a single population splits into two species.

Closely related organisms also develop along similar lines. In general, the paths that they follow diverge dramatically only at the end of their development. They start off at the same place, but, over time, as they are modified by adaptation, they exhibit more differences, although they still retain their underlying genetic and structural identity.

Because they have evolved in this way, organisms pass through the early stages of development that their ancestors passed through, up to the point of divergence. That means that an organism's development mimics that of its ancestors, although it doesn't re-create it exactly. Take the example of a flatfish called Pleuronectes. Early in its development, the fish develops a tail that comes to a point. In the next stage of development, the top lobe of the tail becomes larger than the bottom lobe. However, when development is completed, the two lobes are equal in size. This developmental pattern mirrors the evolutionary transitions that the fish has undergone. At one stage in evolution, Pleuronectes had a tail with a point; at another stage, it had a tail with a top lobe that was larger than the bottom one.

This is what the German philosopher Wilhelm Haeckel was referring to when he said that "ontogeny recapitulates phylogeny" (ontogeny is embryonic development, and phylogeny is evolutionary development). In birds, reptiles, and mammals—vertebrates all—the pattern of cell growth in the early stages of embryonic development is similar, although there are superficial differences in appearance. All these vertebrates reach a primitive, fish-like stage within a few days of fertilization. From that point on, however, development diverges dramatically. There are many exceptions, though; evolution doesn't always recapitulate ancient forms.

Vestiges of the Great-Great-Great-Grandfather

Sometimes traces of an organism's ancestry can be seen even when it has completed its development. These are called vestigial structures. Many snakes, for instance, possess rudimentary pelvic bones that they have no need of—the bones are souvenirs that come from their ambulatory ancestors. When we refer to a vestigial structure, we don't mean that it's useless, only that it is clearly a vestige of a structure inherited from ancestral organisms. Vestigial structures may be co-opted to perform new functions. In humans, for instance, the appendix has become a repository for some immune system cells. Many scientists are convinced that the appendix once served some important function during the course of human evolution, but they are unable to say what exactly it was.

Natural selection has only so much material to work with, of course. Genetic change has its limits, as we've said before. A turtle does not develop a steel shell, however practical such an addition would be, because its genetic inheritance can stretch only so far. In addition, natural selection acts with no foresight. The only adaptation that counts is the one that is pertinent for the present circumstances. Scientists predicted that, as a result, natural selection would have produced some jerry-rigged solutions to a number of design problems in living species. In other words, evidence of common ancestors could be found by examining odd and unusual structural or behavioral patterns. In fact, this is what happens.

For instance, female lizards belonging to the genus Cnemodophorus reproduce parthenogenetically—they don't need males to reproduce. But to increase fertility, these lizards mount other females and simulate copulation. That's because the ancestors of the lizards used to reproduce sexually. Why they gave up heterosexual behavior is unknown. The same hormonal arousal takes place in the descendants—making them more fertile—only the sexual mode of reproduction has been lost. So, nature has contrived an improvised solution that accomplishes the same objective.

Reading the Fossils

Fossil evidence is also used, with considerable success, to demonstrate close relationships between existing organisms and distant ancestors. In rocks, fossils reveal structures that bear less similarity to modern organisms the deeper you dig. The distribution of fossils can tell paleontologists a great deal about where ancient species lived and how far they may have been dispersed. *Radiocarbon dating* indicates that the Earth was formed about four and a half billion years ago. The earliest fossils so far found—microorganisms resembling bacteria and blue-green algae—appear in rocks about three and a half billion year old.

The oldest animal fossils don't show up until relatively recently in geological terms—about 700 million years ago. It was a pretty humble debut, too: The first animals were small, worm-like creatures with soft bodies. It's only when you reach a period of about 570 million years ago that you discover numerous fossils that appear to be the ancestors of many living phyla. These organisms are strikingly different from their present-day descendants, as well as from intervening organisms that they gave rise to somewhere along the evolutionary path. Some of these ancient organisms are so radically different, in fact, that paleontologists had to create new phyla to classify them. Then 400 million years ago, the first vertebrates appeared, but it wasn't until more than 200 million years after that that mammals began to walk the Earth. The earliest humans, on the other hand, didn't arrive on the scene until a little less than two million years ago.

What Does It Mean?

Radiocarbon dating, first developed in 1947, is used to determine the approximate ages of extinct organisms, artifacts, rock formations, and so on by measuring the amount of carbon-14 that they contain. The method is possible because carbon-14 disintegrates at a known rate.

The fossil record is incomplete. Actually, the opposite is the case: Only a very small proportion of organisms were preserved as fossils, and, of those, only a tiny fraction have been recovered and studied. Nonetheless, enough fossils have been found to reveal how forms of organisms evolved over time. The horse, for instance, began its evolutionary journey about 50 million years ago as an animal the size of a dog, with several toes on each foot and teeth suitable for browsing—grazing on leaves. Its modern-day descendant, Equus, is much larger in size, has only one toe on each foot, and has teeth appropriate for grazing. The fossil record preserves a number of transitional forms, including many other kinds of extinct horses that split off in a different evolutionary direction and disappeared altogether, leaving no living descendants. Paleontologists have also shown how traits that developed in transitional forms of life assumed new purposes in different species. For instance, a bone structure in the lower jaw of reptiles evolved into bones now found in the mammalian ear.

Incumbency: Getting in Before the Crowd

One of the most remarkable discoveries that has emerged from the fossil record is the astonishing pattern of development in macroevolution. Families, phyla, and classes didn't come into being at random—far from it. Around 540 million years ago, in a feverish burst of evolutionary activity known as the Cambrian Explosion, all but one of the living marine phyla and many of the marine classes made their sudden appearance, now preserved in the fossil record. Afterward, the pace of both morphological divergence and the production of higher taxa fell off dramatically. Similar evolutionary bursts have occurred among vascular plants and in insects. One of the major challenges facing macroevolutionary theorists was to determine how environmental changes might have promoted and then dampened these evolutionary explosions.

Many of the bursts of evolutionary activity after the Cambrian Explosion all appear to have occurred in the wake of mass extinctions. This observation has led to what macroevolutionists call the hypothesis of incumbency. If you get there first, you are an evolutionary winner—at least, for as long as you can hold the fort. The earlier tenants of a particular habitat remain dominant until they are wiped out or suffer some other calamity. Their eviction then clears the way for other groups to take their place.

The Timing and Tempo of Evolution

The realization that evolution seems to proceed at a particular pace, picking up speed at one point and slowing to a crawl at another, presented scientists with new mysteries to ponder. When they examined the fossil record, they found that many species appeared to be almost static, morphologically speaking, after their first appearance. Why, scientists wondered, weren't these species evolving continuously?

More puzzlingly, the fossil record shows that certain species remained stable for long periods of time and then underwent rapid change—without exhibiting any transitional evolution. In other words, some lineages didn't gradually emerge after splitting off from their parent populations, the way classical textbooks said was supposed to happen. They seemed to go from A to C, without passing through B. How could we account for all these different evolutionary tempos and modes? Did they occur at the same rate among *taxa* and in different habitats, regions, and ecological niches?

Maybe, biologists speculated, evolutionary processes operated simultaneously at several different levels. Factors that might accelerate speciation rates within one population might shape the long-term evolution of a larger group. As you might imagine, trying to sort through all these influences

What Does It Mean?

Taxa is a general biological classification arranged in a hierarchical order from kingdom to subspecies. A taxon (singular) ordinarily includes several taxa of lower rank.

taking place among many different groups and species in different environments at different times continues to provide a great deal of employment for paleontologists, evolutionary biologists, and molecular researchers.

The Truth About Cats and Dogs

There is no significant difference between micro- and macroevolution except that, in microevolution, genes usually combine, and, in macroevolution, genes between species usually diverge. In the former case, offspring are produced; in the latter case, species are produced. The same processes that cause evolution to take place within an individual species are responsible for evolution that takes place in groups above the species level—among taxa, families, and classes, and so on. The origin of higher taxa does not require some special mechanism. The two ancient forbears of canines and felines, for example, probably differed very little from their common ancestors or from each other. But once the ancient cats and dogs became reproductively isolated from each other, they evolved into more differentiated species. This is true of all lineages, back to the first single-cell organisms that mark the birth of life on Earth.

The Least You Need to Know

➤ Macroevolution considers the evolutionary processes above the species level, and microevolution considers evolutionary processes below the species level.

➤ The same processes that can account for microevolution can also explain macroevolution.

➤ Microevolutionary theory, developed in the 1930s, says that existing species all descended from common ancestors.

➤ Evidence for common ancestors of present-day lineages comes from studying closely related species.

➤ A close relationship can be established by similarities of structure and gene sequence, the fossil record, and geographical proximity.

➤ Fossil evidence indicates that evolution proceeded at a different pace at different times; long periods without any evolutionary change in a species can be followed by bursts of rapid change.

Chapter 17

The Role of Extinction in Evolution

In This Chapter

➤ The environmental factors involved in macroevolution: biotic and abiotic

➤ The role of extinction in evolution

➤ Ordinary and mass extinctions

➤ Extinction or branching: which matters more?

➤ Punctuated equilibrium vs. gradualism

➤ Explaining the absence of transitional life forms

The story of how species are created cannot be separated from the story of how they are destroyed. You can make a good case—and many biologists have—that extinction is an integral part of evolution, that you couldn't have one without the other. Extinction is the yin to evolution's yang. Extinction is a natural and, sadly, inevitable process that occurs every day, more often undetected than not, as species of animals and plants vanish forever from the planet. If anything, the pace of extinction has sped up enormously, most of it attributable to human intervention. Industrialization, pollution, global warming, agriculture, and deforesting are exacting an ever-greater toll on the millions of species that cohabit the Earth with us.

Unlike ordinary day-to-day extinctions, mass extinctions occur only once in a great while. There have been five known mass extinctions, the last of which occurred about 65 million years ago. The extermination of the dinosaurs occurred as a result of such a mass extinction, although its cause is still disputed.

When macroevolutionary theorists examined the patterns of extinction and speciation—the creation of new species—they wondered whether extinction was a result of evolution or was, as proponents of a theory called punctuated equilibrium contend, a force that drives evolution. It's a classic chicken-or-egg dilemma. Certainly, mass extinctions appear to propel evolutionary change into high gear by uprooting long-established dominant populations (such as dinosaurs) and opening the way for other, less privileged populations (such as mammals) to usurp their place. In reexamining the role of extinction in evolution, punctuated equilibrium, whose leading advocate is Harvard paleontologist Stephen Jay Gould, also called into question the rate at which evolution takes place.

What Does It Mean?

Biotic factors refer to the influence of other living organisms on an organism. **Abiotic** factors refer to any physical factors in an environment that influence the organism.

In contrast to the traditional view advanced by Darwinians, punctuated equilibrium, introduced in 1972, states that most evolutionary change doesn't occur gradually over long periods of geological time (meaning millions of years), but takes place in furious bursts lasting no more than 5 to 50,000 years. Gould and other advocates of the theory point to an absence of transitional life forms in the fossil record between earlier species and species that are found in strata just above them. Although many aspects of punctuated equilibrium have since been incorporated into Darwinian theory, punctuated equilibrium remains controversial and far from fully proven.

Shaping the Patterns of Change

When evolutionary biologists study evolution, they have to take into account a bewildering range of factors, any of which can shape macroevolutionary patterns by affecting adaptive success or the likelihood of survival. Biologists refer to two basic categories of factors that need to be considered—*abiotic* and *biotic*. Abiotic factors refer to physical factors in an organism's environment—water, climate, light, oxygen, and so on. Biotic refers to other living organisms—predators, parasites, prey, and so on—that have an influence over the organism.

What Does It Mean?

An **ecosystem** refers to the organisms living in a particular environment—jungle, sea, forest—and the physical components of an environment that affect them. The term was first used by British ecologist Sir Arthur George Tansley in 1935.

All interacting biotic and abiotic factors, taken together, make up an *ecosystem*. Organisms are constantly interacting with their environment; at thit operates like an endless feedback system. These environmental factors ultimately check the growth of a given population because, sooner or later, it will use up its resources, be driven out by competitors, or experience some environmental calamity. (It was Darwin's insight to see that, although a population could grow geometrically, as Malthus had predicted, its expansion ultimately would be limited by nature.) The limit to a population's growth is called a carrying capacity.

The respective roles that physical and biotic factors have in influencing macroevolutionary development remain a subject of fierce debate among scientists. The growth of a population will be influenced to a great extent by its genetic fitness and its ability to ward off predators or withstand environmental stresses—all of which fall under the category of biotic factors. Incumbency, too, is a biotic factor. As you'll recall from the last chapter, incumbency refers to a population commanding a particular environmental niche before competitors can get there. It's the "firstest with the mostest" idea, a pre-emptive strike that gives the earlier population an advantage over latecomers. Changes in climate, geography, and even atmospheric or ocean chemistry can disrupt the stability of a population—all abiotic or physical factors. Many scientists believe that it was just such a physical factor—on a vast scale—that caused the mass extinction of dinosaurs at the end of the Mesozoic Era (around 65 million years ago).

Some Species Die So That Others Might Live

For macroevolutionists, the role of extinction, particularly mass extinctions such as those that carried off the dinosaurs, has been a major preoccupation for years. Do peaks in extinction merely accelerate processes already set in motion during times of relative stability, or do they play a special role all their own? Some evolutionary biologists point out that mass extinctions appear to break the dominance of incumbent groups, thereby opening up opportunities for new species to flourish and spread. This would lead to diversification and the development of new evolutionary forms. Until the dinosaurs perished, for example, mammals could barely make much headway. Now, as we are all well aware, mammals enjoy domination over the Earth. Although most extinctions are not so dramatic—species go under often without making much of a fuss at all—mass extinctions may have had a disproportionate evolutionary impact over the past half-billion years or so.

Most of the history of multicellular life on Earth has been characterized by ecological and evolutionary stability. Dominant forms of marine taxa, for instance, appear to have fairly predictable rates of origination and extinction. Why should marine diversity at the family level, for example, hold steady for most of the Paleozoic when the Cambrian Explosion took place? In other words, on certain levels—a family, say, or a class or a species—stability is the norm; at the same time, on higher levels (among taxa), all hell is breaking loose and populations are being wiped out right and left. Does this suggest that there might be some kind of a global equilibrium—a debit-credit system governing the extent of diversity and novelty—where turmoil and turnover on one level is offset by stability on another? As yet, no one can say for sure.

Ordinary Extinction: Obits Hardly Anyone Reads

Before we go on, we should consider what extinction means. Sad to say, extinction is the ultimate fate of all species—and, as far as we know, there's no exemption for humans. Extinction can occur for many reasons. Species can be pushed out of their habitat by competitors, or the habitat itself might disappear—a lake dries up, or a flood submerges a forest. Sometimes the organisms that the species have been thriving on for generations develop an unbeatable defense, depriving the species of its food source. Some species enjoy a long stay on Earth, while others have very brief life spans. Why that should be is unknown.

Although some biologists believe that species are programmed to go extinct, most believe that species can survive indefinitely as long as their environment remains relatively stable. So, we can look at extinction in two ways. One way is to see extinction as the result of a natural process, in which species originate, develop, exploit the resources of a particular habitat, and eventually are eliminated from the running by a variety of environmental factors. But extinction can also be viewed as a cause of evolutionary change, the spark that triggers a sequence of events leading to more diversity and novelty.

Imagine a mass extinction occurring today, with only a few life forms surviving. Over millions of years, those survivors would give rise to other types of life forms. Given a lot of time—say, a billion years—life probably would look little like it does today. The diversity and novelty would be inconceivable to us. Taxonomists in that remote future would have to develop a whole new system to classify the life forms that arose.

Mass Extinction: Cleaning the Slate

If we picture evolution as a branching tree, it's one that has been severely pruned several times in its life. Species are becoming extinct even today (most as a result of human intervention). Some branches were cut suddenly before they had a chance to grow very much—evolutionary dead ends. As we mentioned earlier, some earlier species of horses occupied a branch all their own and simply disappeared, leaving behind no traces in the horses that we know today. Mass extinctions are relatively rare, but when they do occur, their impact is enormous. Thousands of species are destroyed in one fell swoop. These extinctions are followed by periods in which new species evolve furiously to fill in the empty ecological niches left behind. These species might then *radiate*—spread out in all directions from the original niche.

For the surviving organisms, a catastrophe such as a mass extinction is a stroke of luck. Thus, in that respect, we can say that contingency or chance plays a significant role in patterns of macroevolution. The rules of the game are changed without warning, and species that might have been losers, or those barely keeping up, under the old rules now find themselves the winners—and sometimes the only ones left at the table.

Memorable Mass Extinctions

The largest mass extinction is believed to have taken place about 250 million years ago, at the end of the *Permian Period*. It was a time of unprecedented turbulence. Continents were being raised from beneath the shallow seas, and mountain ranges—including the Appalachians in the United States and the Urals in Russia—came into being, unceremoniously thrust into existence by intense geological pressures. Europe was joined to Asia, and North America was shoved into a vast land mass that formed a supercontinent called Gondwanaland. (The explanation of how all these upheavals came about was incorporated in the theory of continental drift, proposed in the early 1900s by Alfred Wegener. We'll discuss continental drift and plate tectonics, its modern-day successor, in more detail in the next chapter.)

Inevitably, these cataclysmic changes had a profound effect on organic life by destroying vast numbers of species and producing new opportunities for the survivors. Invertebrate marine life at the beginning of the period flourished in the warm, shallow inland seas, but by the time the dust had settled several million years later, most of them were gone. On land, thousands of species of seed ferns, conifers, and ginkgoes were eliminated. And even as amphibians were declining in number, the mass extinction was a boon to reptiles, which were poised to enjoy their day in the sun. It was at the end of this tumultuous era that the forerunners of the dinosaurs first made their appearance.

What Does It Mean?

The **Permian Period** is the last division of the Paleozoic era and occurred between 290 million and 240 million years ago. It was marked by a period of great upheaval in which continents were formed and many of the world's great mountain ranges arose.

The dinosaurs had nearly 200 million years to rule the roost. As the dominant species nowadays, we still have a long way to go to catch up to their record. But then, around 65 million years ago, somewhere between the Cretaceous and Tertiary Periods (known to geologists as the K/T Boundary), it was the dinosaurs' turn to face the music. Some scientists believe that the dinosaurs were wiped out by a cascade of events set off by an asteroid crashing into Earth in present-day Mexico. According to proponents of this theory—and it is one that is hotly disputed—the impact of the asteroid triggered a fireball and caused debris and ash to accumulate in the atmosphere, forming a cloud of dust that blocked sunlight for several months. So, in addition to the devastating effects of the initial fire, the long-term effects of the fallout were even worse, producing conditions similar to what might occur in the aftermath of a nuclear war. The rain, for instance, was intensely acidic, filled with scorched sulfur from the site of the asteroid's impact and nitrogen present in the air.

Hit with a double whammy—the darkness combined with toxic rainfall—plants withered and died. Herbivorous dinosaurs, which fed on plants, succumbed in turn, but so did the carnivorous dinosaurs, which depended for their food on organisms that lived off plants. Before the extinction, according to this scenario, mammals lived in the shadows of the dinosaurs, mostly subsisting on insects. But the cataclysm that wiped out the food sources for the dinosaurs provided an unprecedented feast in the form of decaying plants for frogs, lizards, turtles, and mammals, allowing them the freedom to develop in the ecological niches that had suddenly opened up.

Other biologists discount the asteroid theory, preferring a far less sensational explanation for the fate of the dinosaurs. Gradual environmental change, they contend, marked by shallow seas receding from the continents, kept putting added pressure on the dinosaurs, removing their habitats and food sources little by little over the course of million of years, until they died out entirely. Whatever theory is correct, though, the result is not in dispute: The mammals were the clear winners.

Groups of mammals, too, have undergone periods of mass extinction at various intervals, though nothing comparable to the dinosaurs. (After all, had the equivalent happened to mammals, none would exist today and you wouldn't be here reading these words.)

Altogether, the paleontological evidence indicates that there have been five great mass extinctions on Earth. In each of these events, between one fourth and one half of all species were wiped out within a course of a few million years. Now many scientists are warning of a sixth. Experts now say that, in this century, it's possible that between one third and two thirds of all plant and animal species, most of them in the tropics, will be lost. If a mass extinction of this scale occurs, it will also take place far more rapidly—in hundreds, not millions, of years. And this time there will be no uncertainty about the cause: The loss of species will come about directly because of the population explosion.

What Does It Mean?

The **background rate of extinction** refers to the average rate of extinction since the last mass extinction 65 million years ago. The background rate gives scientists a way of judging whether the rate of extinction is increasing or decreasing.

Animal and plant life is being purged by a combination of agriculture, deforestation, and the exploitation of natural resources, to say nothing of air and water pollution or the increasing carbon emissions that contribute to global warming. Humans have been killing off species, intentionally or unwittingly, for several thousand years, so the phenomenon is nothing new. But because of ever-greater numbers of people competing for resources, especially in the developing countries of Africa and Asia and more intensified industrialization in the developed countries, the destruction is accelerating.

In making their projections of future extinction, scientists rely on what's called the *background extinction rate*, which measures the approximate rate at which species have been becoming extinct over the last 65 million years since the dinosaurs were killed off. The current extinction rate is now close to 1,000 times that of the background rate, meaning that species are now disappearing far faster than they have on average over the course of many millions of years. In this century, if current trends continue, the rate of extinction may climb to 10,000 *times* the background rate!

Punctuated Equilibrium

The idea that evolution proceeds by small steps and never makes jumps is a key principle of neo-Darwinism. It is called *gradualism*. Darwin supposed that variation would involve very small changes that would add up over time and result in the creation of new species. The most significant challenges to gradualism comes from Stephen Jay Gould, a professor of geology and paleontology at Harvard University, and Niles Eldredge, of the American Museum of Natural History. In a paper published in 1972, they proposed that evolution proceeds in fits and starts. The two cited the fossil record to argue that populations can enjoy long periods of stability (or stasis) during which virtually no evolution occurs, but that these periods were "punctuated" by relatively short periods of rapid evolution. Periods of stasis, the two contended, could last for up to several million years, while the bursts of evolutionary activity would take only 5,000 to 50,000 years—a blink of an eye in geological time.

Punctuated equilibrium doesn't say that some evolutionary change isn't gradual. Instead, it states that "most" evolutionary change happens in episodes that are geologically very brief. The theory of punctuated equilibrium is an inference about the process of macroevolution based on the fossil evidence. While punctuated equilibrium has more or less been integrated into neo-Darwinism, many evolutionary biologists who are not convinced that the theory will hold up.

What Does It Mean?

Gradualism is a concept of Darwinian theory that states that new species are created as a result of gradual, though not necessarily uniform, accumulation of many small genetic changes over long periods of time.

What the Fossil Record Shows

What was it about the fossil record that led Gould and Eldredge to propose their theory in the first place? If the gradualists were correct, they said, the fossil record should show a progressive development of species from one form to another over time, as preserved in strata of rock. But with few exceptions, this isn't what paleontologists have found. Instead, the fossil record appeared to suggest that the transition from one species to another is usually abrupt in most locales. That convinced Gould and Eldredge that species remain unchanged for long stretches of time and then something happens—an extinction or sudden change of climate—and a new species emerges. These new forms then invade the territory previously occupied by their ancestral species. That doesn't imply that the new species evolved from the older one. Instead, the new species migrated from elsewhere when a suitable ecological niche was vacated by the earlier species.

We should point out, however, that, in some cases, transitional forms of life have been discovered that bridge the gap between an earlier and a later species. In one location, for instance, fossils of two species of the phyla brachiopod (which includes mollusks such as the lampshell) were discovered lying in strata separated by a narrow layer of 10 centimeters (about an inch). In addition to the two species, transitional forms were unearthed along with them.

The course of evolutionary change over the last half-billion years, Gould believes, has been far more influenced by extinction than by a proliferation of branches on the tree of life. We return to the question posed earlier: Is extinction a driving force of evolution, or is it the reverse, extinction as a result of evolutionary forces? Or, depending on the time and the circumstances, is it both?

Contingency vs. Predetermination

Punctuated equilibrium emphasizes the importance of contingency in macroevolution. In other words, evolution was not predetermined. It did not have an objective; if the evolutionary tape were rerun and started over again, this time with the circumstances changed over the course of 500 million years, the results would be quite different than what we see today. If mass extinctions hadn't occurred, or if they had occurred at another time than they did, organic life could have developed in an entirely different direction. In that case, Gould said, humans would have been unlikely to appear at all.

Some critics of Gould's theories argue that, on the contrary, evolutionary branching is a more widespread phenomenon than Gould credits. In their view, innovation, not extinction, is seen as the major driving force of evolution. New and more complex life forms will tend to develop, regardless of what course evolution had taken. Sooner or later, even given an alternative scenario of events, inevitably a complex life form like humans would have emerged.

Species Selection: Why the Controversy?

The most controversial idea proposed by punctuated equilibrium is called species selection. Advocates of the theory draw an analogy between mutation and speciation. That is, just as mutations add new alleles to the gene pool, speciation adds new species to the species pool—thus, the term *species selection*. Species selection, then, is natural selection operating on another level. Natural selection chooses favorable alleles over unfavorable ones. Similarly, species selection chooses better fit species over lesser fit ones. Does this really happen? Of all the ideas that make up the theory, species selection is certainly the most controversial. Even biologists who go along with many aspects of the punctuated equilibrium theory contend that species selection isn't even theoretically possible and scoff at attempts to relate it to natural selection. Saying that the processes that generate alleles and species are the same is, in their view, comparing apples and oranges.

Who Are They?

Stephen Jay Gould (1941–) is a noted professor of geology and paleontology at Harvard University who has written extensively about evolution. Together with Niles Eldredge, of the American Museum of Natural History, he originated the theory of punctuated equilibrium in 1972, based on the fact that the fossil record contains few transitional forms of life.

Why Punctuated Evolution Gets So Much Attention

Most modern evolutionary theories aren't much discussed outside the realm of academia. But punctuated equilibrium is a notable exception. That so many people are aware of it is largely due to the popularity of Gould's best-selling books and monthly articles in *Natural History* magazine. Unlike many of his colleagues, Gould writes for the general public, so he has a decided advantage over some of his challengers who confine their arguments to scientific journals. Whether punctuated equilibrium can ever be proven correct, in whole or in part, is still unclear, and the debate over its validity is likely to continue for a very long time.

The following principles form the basis of the theory of punctuated equilibrium:

➤ As a theory, punctuated equilibrium was first proposed in a 1972 paper by Stephen Jay Gould of Harvard University and Niles Eldredge of the American Museum of Natural History.

➤ Species do not evolve for long periods of time, often extending for millions of years (when they are said to be in stasis), but they undergo rapid change in a short geological span of time of 5,000–50,000 years.

➤ Punctuated equilibrium suggests that evolution proceeds in leaps rather than gradually over time as a result of an accumulation of small changes. This view was accepted for a long time by neo-Darwinians.

➤ Gould and Eldredge based their case on the fossil record, which reveals few transitional forms bridging the gap between earlier species and later ones. The two concluded that the absence of such forms was an indication that one species could die out and another could emerge to replace it without any intermediate evolutionary steps leading from one to the other.

➤ Punctuated equilibrium sees extinction as a more powerful driving force of evolution than the creation of new life forms.

➤ Contingency or chance plays an important role in evolutionary change. The emergence of the human (or any other modern) species was not inevitable. A different pattern of evolutionary and environmental change might have led to entirely different forms of life than those existing today.

➤ The most controversial aspect of punctuated equilibrium theory holds that a process known as species selection is taking place on a macroevolutionary level that is analogous to the addition of new alleles to a gene pool by natural selection.

➤ Many aspects of punctuated equilibrium have been incorporated into modern Darwinian theory, but the theory is still vigorously disputed by many scientists and has yet to be proven.

The Least You Need to Know

➤ Many environmental influences must be taken into account in macroevolution to determine the rate of evolutionary change and the way in which change affects populations at various levels: taxa, family, class, species, and so on.

➤ Evolutionary change is inextricably tied in with extinction.

➤ Extinctions happen in two basic ways: gradually and routinely, or in mass extinctions that eliminate a quarter to a half of all existing species in a relatively short geological period—up to five million years.

➤ Five mass extinctions have taken place in the Earth's history, with the last one causing the disappearance of the dinosaurs about 65 million years ago.

➤ Whether extinction or the creation of species is a more important phenomenon in evolution is still a subject of dispute.

➤ The theory of punctuated equilibrium, first proposed by Stephen Jay Gould of Harvard and Niles Eldredge of the American Museum of Natural History in 1972, asserts that extinction is a more important force than speciation.

Chapter 18

Continental Drift and Evolution

In This Chapter

➤ Explaining Earth's shape by catastrophe

➤ How scientists determined the age of the Earth

➤ The theory of continental drift

➤ How drift and plate tectonics may explain evolutionary change

In the seventeenth and eighteenth centuries, geology was still dominated by the concept of the biblical Flood as a major force in shaping the Earth's surface. The prevailing theory of the time was catastrophism—the belief that environmental calamities, such as the Flood, had brought about sudden and radical geological changes. In 1756, a German theologian named Theodor Christoph Lilienthal concluded from biblical references that the Flood must have ripped apart the surface of the Earth. As evidence of his position, he pointed out how well the coastlines of eastern South America and western Africa appeared to fit together.

But it would take several centuries before anyone would come up with a compelling explanation to account for why the two continents looked as if they belonged together. The credit belongs not to a geologist, but to a German astronomer and meteorologist named Alfred Wegener. And the reason he gave was that appearances weren't deceiving; Africa and South America looked like they fit together because they did.

As we've seen earlier in this saga, no real understanding of how evolution proceeded on Earth was possible without an understanding of geology. The two subjects were inextricably linked. You couldn't begin to figure out how life developed on Earth if you had no idea how old the Earth actually was. But that was only a first, albeit major, step. The Earth's surface, geologists discovered, was no more static than species were. Over eons, the crust of the planet had been subjected to countless upheavals. A map of the world, drawn a few hundred million years ago, would look nothing like a map of the world today.

And as the shape of the Earth changed, as mountain ranges were born and ground down by erosion, and as land sunk and seas swept over them, only to recede millions of years later, organic life was changed, in turn. Habitats in which a population had flourished were suddenly erased, while new ones, eminently suitable for a new, better adapted population, came into being. But, as you might expect, efforts to piece together the geological puzzle of the Earth's history—ironically, by disassembling the pieces—was as daunting a challenge as determining how species evolved and disappeared.

From Earth to Moon and Back Again

Early in the nineteenth century, the German naturalist Alexander von Humboldt proposed that the Atlantic had been carved out by a catastrophe that formed a gigantic "valley" with parallel borders. Then in 1858, a catastrophist named Antonio Snider-Pellegrini came up with another explanation for the existence of the Atlantic. He suggested that the Flood (always a convenient suspect) had caused material to pour out of the interior of the Earth and that this expelled material had caused the rift between South America and Africa, occupied by the Atlantic.

Who Are They?

Alexander von Humboldt (1769–1859) was a German naturalist and explorer who contributed to a deeper understanding of geophysics, meteorology, and oceanography.

The similarity in shape between the two coastlines wasn't the only argument in favor of his theory. Snider-Pellegrini also pointed out that there were unusual similarities in the fossils and certain rock formations on opposing sides of the ocean. He subsequently published a map in which he reunited the Americas, Europe, and Africa, foreshadowing the idea that a supercontinent once might have existed.

Other theorists offered more novel explanations for the creation of the Atlantic. The English astronomer Sir George Darwin (Charles's second son) proposed that the moon had been thrown off from a fast-spinning Earth early in its history. The result was a gaping hole in the Earth's surface previously occupied by the Moon-making material, said Darwin, and it was soon filled by what is today the Pacific Ocean. The same process that created the Pacific dragged the Americas away from Europe and Africa. His Moon theory never got off the ground.

A Record of Violence

Toward the end of the nineteenth century and the beginning of the twentieth century, catastrophist notions began to fall out of fashion, replaced with the idea of uniformitarianism, favored by the geologist Charles Lyell. As you may recall, uniformitarianism said that all the processes observable in nature today were always at work even millions of years ago. Certainly catastrophes occur, but why should they have been so frequent in the past, as catastrophists claimed, and so rare at present? To advocates of uniformitarianism, there had to be a more plausible explanation.

Who Are They?

George Darwin (1845–1912) was an astronomer and the second son of Charles Darwin. He became best known for his studies of the tidal motion and the Earth-Moon system, postulating that the Moon was formed from material that originated in a young, still molten Earth.

Yet uniformitarianism had trouble accounting for the growing fossil record. In fact, the evidence from the fossils appeared to suggest that the geological forces active on Earth were a great deal more violent and the geological changes were far more dramatic in the distant past than they are now. Polar areas now draped in ice, for instance, once nourished hot, tropical climates.

The Earth's Age

As some geologists tackled the problem of deriving an accurate picture of the Earth's surface in the ancient past, others were struggling to calculate the planet's true age. The key to a solution seemed to lie in the chemical composition of the Earth's interior. It was hot—that much everyone agreed on. But what exactly was burning and how long had the fire been going? If they could answer these two questions, geologists believed that they might be in a position to establish the age of the Earth.

Early in the nineteenth century, the English chemist Sir Humphrey Davy proposed the idea of continuous "fires" raging down below, caused by oxidation processes. This theory left something to be desired, though. Later in the century, Lord Kelvin (William Thomson) calculated that Davy could not possibly be correct. Even if at its birth

Who Are They?

Sir Humphrey Davy (1778–1828) was a pioneering British chemist who was the first to understand that electricity in simple electrolytic cells was caused by the chemical action of two substances with opposite charge. He also was the first to isolate potassium, sodium, boron, and the alkaline earth metals from their compounds.

the Earth had been as hot as the surface of the sun, Kelvin pointed out, it could not be more than a few hundred million years old and still have as much internal heat left as scientists had observed. That presented a dilemma for geologists. If the Earth were only a few hundred million years old, then how could they explain the age of the rock formations that clearly were far older? Nor was such a time span sufficient to explain how great mountain ranges had been repeatedly built up and eroded away. Whatever processes were generating heat in the Earth's interior, they had to have been going on for far longer.

It was only at the end of the nineteenth century that the discovery of radioactivity resolved the dilemma as to how the Earth's interior was kept heated. Radioactive elements—mainly uranium, thorium, and a radioactive form of potassium—are found in many kinds of rocks. These elements undergo slow decay that transforms them into other elements, a process that generates heat. While the heat on the Earth's surface dissipates so rapidly that it is virtually unobservable, it accumulates deep in the Earth, where it provides a continuous source of energy.

DARWIN

Who Are They?

William Thompson (Lord) Kelvin (1824–1907) was a British mathematician and physicist who investigated fields ranging from chemistry and magnetism to electricity and energy. In 1842, he proposed a measurement of absolute temperature that bears his name today.

Pulling Apart the Supercontinent

At this point Alfred Wegener enters history. Born in Berlin in 1880, he represents one of the last of the adventurer-naturalists who dominated the history of science in the eighteenth and nineteenth centuries. Although he received his Ph.D in astronomy, he almost immediately turned his attention to meteorology, then a new science. Throughout his career, Wegener was drawn by curiosity to fields as diverse as paleontology, oceanography, seismology, geomagnetism—and, yes, evolution. As a result, he had a vast and far-ranging body of knowledge to draw upon. But by willfully violating rigid academic and professional boundaries and poaching on territory that others considered their own preserve, he didn't make things easier for himself or for the literally earth-shattering theory that he presented to the world.

The Debut of Continental Drift

Like von Humboldt, Wegener was struck by the fit of the African and South American coastlines. In 1912, even before he had carried out any scientific investigations, he first put forward the idea of "continental displacement" or what later was called *continental drift*. The inspiration for continental drift was sparked by a scientific paper that he came across while browsing in the library of the University of Marburg in Germany. The paper documented the discovery of fossils of identical plants and animals on opposite sides of the Atlantic. Intrigued, Wegener began to search for other cases in which similar organisms were separated by great oceans.

At this time, geologists believed that the continents hadn't split off from one another. Instead, they held that the continents had been connected by land bridges but that, over the years, those links had been submerged under water. Wegener was unconvinced. The similarities among the organisms, just like the fit of the coastlines, he maintained, were a consequence of the separation of great land masses that had been joined together at one time. The hypothetical land bridges, he insisted, couldn't be found because they had never existed. As he later wrote, "A conviction of the fundamental soundness of the idea (of continental drift) took root in my mind."

Wegener's pursuit of the theory of drift had to be put on hold when he was called on to serve at the front at the outbreak of World War I. Sidelined by a wound that he received on the battlefield, he took advantage of his prolonged convalescence to resume his research. Confident enough that he had assembled enough facts in support of his position, he published the first comprehensively theory of continental drift, *The Origin of Continents and Oceans.* (The book is now considered a classic.) While Wegener wasn't the first to suggest that the continents had once been connected, he was the first to marshal extensive evidence to back it up from several diverse fields.

Wegener contended in his book that large-scale geological features on separated continents often matched very closely. For example, the Appalachian Mountains of eastern North America shared many of the same features of the Scottish highlands, and the distinctive rock strata of the Karroo system of South Africa were identical to those of the Santa Catarina system in Brazil. Wegener also drew on fossil discoveries to support his theory. Parts of the globe that had once basked in tropical warmth had relocated to frigid climes. Excavations on the Arctic island of Spitsbergen, for instance, had revealed fossils of tropical plants, such as ferns and cycads. South Africa was once partially covered by ice, based on evidence from finds of deposits of mixed sand, gravel, boulders and clay, which typically are left behind by melting ice sheets. Similar phenomena could be observed elsewhere.

What Does It Mean?

Continental drift is a theory proposed by the German scientist Alfred Wegener stating that all present-day continents once composed a single land mass that began to break apart in successive stages and drift away from each other, eventually reaching their current locations. The process is still continuing, albeit almost imperceptibly.

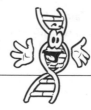

A Natural Selection

The hippopotamus was quite a nomad early in its history. Fossils of the animal have been unearthed in European and Indian deposits of the Pliocene Epoch (5 million to 1.5 million years ago). Similar fossils found in England seem to be of the same species as the present-day hippopotamus.

Each region of the Earth, in the past as well as now, has tended to have its characteristic vegetation. All parts of the world have their characteristic vegetation. Drop an experienced botanist down anywhere on Earth, and he should be able to tell you, based on the local vegetation, where he is, give or take a few hundred miles. But 200 million years ago or so, a hypothetical botanist would have been unable to make such a distinction if the only evidence that he had to go on was a seed fern, known as Glossopteris. Fossil remains of Glossopteris, which flourished during the Carboniferous Period (360 million to 290 million years ago), have turned up all over the place and in such widely separated regions as India, Australia, South America, and South Africa. The famous expeditions early in the twentieth century led by Scott and Shackleton to Antarctica even found the plant embedded in coal seams in mountains near the South Pole. How could one form of vegetation have been common to lands now separated by thousands of miles of ocean? That suggested to Wegener that, at one point, all these land masses had been one and the same.

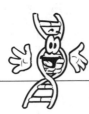

A Natural Selection

Marsupials are a group of mammals in which the females characteristically have pouches for carrying their young. Marsupials include the kangaroo, koala, opossum, numbat, ringtail possum, and Tasmanian wolf. They are mainly found in Australia, New Zealand, and parts of South America.

Or take the example of the lumbering hippopotamus, said Wegener. The hippopotamus is found on Madagascar as well as on mainland Africa, where it seems to have originated. Hippocampuses evolved to wallow in muddy streams. Was it possible that they had ever been such extraordinary swimmers that they could have somehow made the crossing of 250 miles of open sea to reach Madagascar? It didn't seem likely. Wegener believed that the explanation must lie in the fact that, at one point, Madagascar and Africa had been joined. The distribution of marsupials revealed a similar pattern; these mammals are found only in Australia, New Zealand, and South America. That led Wegener to contend that Australia and New Zealand had once been joined to South America.

Gondwanaland: When India and Africa Lived Together

Wegener wasn't alone in observing that India, Australia, New Zealand, and the Antarctic might all have been one. The Austrian geologist Eduard Suess had remarked on the similarity of rock formations—largely consisting of sandstone, with coal, reptilian fossils, and Glossopteris—in India, Africa, and Madagascar. Suess proposed that these southern regions were all linked by a giant continent, which he called Gondwanaland (naming it for a region of India inhabited by an aboriginal tribe, the Gonds, which is rich in the fern-seed plant Glossopteris.) Suess believed—mistakenly, as it turned out—that Gondwanaland was a continent, parts of which had sunk beneath the water. (In proposing this idea, Seuss helped revive the mythical concept of Atlantis.) Wegener, however, was correct to insist that the ancient land masses hadn't "sunk," but had rather fractured and drifted apart.

Pangaea: The Supercontinent

All continents had once been a single land mass that had existed 300 million years ago, Wegener asserted. He called this supercontinent Pangaea (from the Greek for "all the Earth.") Pangaea had split apart, Wegener said, and its pieces had been on the move ever since. Pangaea hadn't broken apart all at once, but had fragmented in a series of ruptures. Antarctica, Australia, India, and Africa began separating in the Jurassic Period—the "Age of Dinosaurs"—some 150 million years ago. In the Cretaceous Period that followed, Africa and South America separated "like pieces of a cracked ice floe," in Wegener's words. The final separation took place at the start of the Ice Age, about a million years ago, and tore off Scandinavia, Greenland, and Canada.

Evidence of this last separation is still seen in the Mid-Atlantic Ridge that rises above the sea in places to form such islands as Iceland and the Azores. As Wegener put it, "It is just as if we were to refit the torn pieces of a newspaper by matching their edges and then check whether the lines of print run smoothly across. If they do, there is nothing left but to conclude that the pieces were, in fact, joined in this way."

The geologists of Wegener's day, still wedded to the idea that the continents had been linked by now-sunken land bridges, scoffed at the theory of continental drift, and Wegener spent the remainder of his life—he died in 1930 on an expedition to Greenland—defending it. Although he was mistaken in some of the details, maintaining, for instance, that the continents were drifting from one another at a faster rate than they actually are, the theory has turned out to be the best model of geological change that scientists have. It wasn't until the 1960s that Wegener's theory was revised and embraced by geologists. Exploration of the deep, beginning in the 1950s, had confirmed what Wegener had predicted—that the chemical composition of the sea floor was different from the composition of the continents. That finding suggested that there couldn't have been any land bridges. Otherwise, scientists reasoned, the composition should have been the same.

Who Are They?

Eduard Suess (1831–1914) was an Austrian geologist whose original theories on geological formation included the idea that the southern continents had once formed a unified land mass, which he named Gondwanaland.

Plate Tectonics and the Vindication of Wegener

Inevitably, continental drift underwent several revisions. The theory that has come to replace it is known as *plate tectonics*. According to plate tectonics, the Earth's surface is made up of a number of large, rigid plates that float on a layer of rock called the

What Does It Mean?

Athenosphere is a layer of Earth made up of rock supporting the rigid tectonic plates that make up the outer surface of the Earth. The relatively malleable consistency of the athenosphere allows it to move and circulate.

athenosphere. Plate tectonics is used to explain a variety of geological processes, including volcanic activity, earthquakes, and the creation of mountains. But the modern theory retains the basic principle of continental drift in holding that the continents today were once part of a single land mass that split off into pieces, all of which have been in motion for hundreds of millions of years.

The theory of continental drift is based on the following principles:

➤ All present day continents were once united in a supercontinent, a single large land mass that existed about 300 million years ago.

➤ The supercontinent gradually broke apart in successive waves, and each piece began to drift apart from the others.

➤ Evidence to support the theory, proposed first in 1912 by the German scientist Alfred Wegener, comes from a variety of sources: the similarity of the west African and eastern South American coastlines, the similarities of widely distributed vegetable fossils, and the similarities of animals in lands separated by large bodies of water.

➤ Scientific advances beginning in the 1950s allowed scientists to refine the theory of continental drift and propose that the outer surface of the Earth is composed of rigid tectonic plates that are in constant motion. Plate tectonic theory, as it's called, accounts for a number of geological phenomena, including seismic and volcanic activity as well as the spread of the sea floor and changes in magnetic polarity throughout the history of the Earth.

What Does It Mean?

Plate tectonics is a unifying theory of geology that was developed in the 1960s and builds on the theory of continental drift. It states that the outer surface of the Earth—the lithosphere—rests on a number of rigid plates that move about on a softer layer called the athenosphere. The theory is used to explain many geological phenomena such as earthquakes, volcanic activity, and mountain building.

Wrenching the Poles: Explaining the Cambrian Explosion

Continental drift and its modern-day elaboration, plate tectonics, may turn out to explain several evolutionary mysteries, including the Cambrian Explosion—the appearance of an astonishing number of species around 540 million years ago. This "explosion" gave rise to all the main animal groups that are alive today. Recent scientific research has yielded some intriguing evidence that, at the beginning of the Cambrian Period, the outer layers of the Earth migrated—like a coat being pulled down—until the places that were previously at the poles were relocated to the equator. Given the fact that at this time most of the species inhabited the ocean, such a forced relocation would have profound implications. Evidence for this upheaval comes from studies of the weak magnetic fields preserved in rocks, which can show the direction of magnetic north at the time of the rock's formation—a phenomenon known as *paleomagnetism*. In this way, scientists can see how the rock (and the continent that carried it) has moved over time.

The uprooting of the poles, according to researchers led by Dr. Joseph Kirschvink of the California Institute of Technology, possibly occurred because of a build-up of mass in one part of the Earth, a possible result of Continental Drift. If the mass had weighed enough, it would have altered the Earth's spin in much the way that putting lead weights on a basketball would. Centrifugal force would act on the Earth just as it would any spinning ball with a lump on its surface; over 15 million years, the underlying mantle would have pulled the poles down to the equator, in the process bringing about repeated changes in the direction of the ocean currents.

What Does It Mean?

Paleomagnetism refers to the capacity of rocks to preserve weak magnetic fields, which allow scientists to determine the direction of magnetic north while the rock was being formed to show how rocks (and the continents) move through time.

The frequent shifts of the water currents, in turn would have disrupted large, stable ecosystems, putting an end to the advantage of incumbency enjoyed by established populations. In this scenario, a number of small ecological niches would have been formed, throwing a lot of populations together that had never before had a chance to encounter one another. As a result, a grand new evolutionary stew would have been created.

The biodiversity of the South American rainforests might remain as evidence of these cataclysmic changes. Although this theory hasn't been proven, it nonetheless offers a vivid illustration of how advances in geology based on continental drift and plate tectonics can illuminate some of the most astonishing—and bewildering—developments in evolution at the same time.

The Least You Need to Know

➤ The natural "fit" of eastern South American and west African coastlines suggested to many naturalists of the seventeenth and eighteenth centuries that the two continents once might have been joined.

➤ Similarities in fossils and of rock formations found in widely separated lands offered further evidence that continents had been joined.

➤ A German meteorologist and astronomer, Alfred Wegener, proposed the theory of continental drift in 1912.

➤ According to Wegener, the continents had drifted apart.

➤ Although Wegener was widely derided by geologists of his time, his theory now provides the basis for modern geological study of the Earth's surface.

➤ The theory of plate tectonics, developed in the 1960s, has elaborated on and refined the theory of drift.

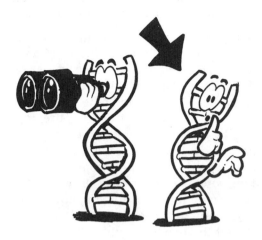

Chapter 19

The Search for the Missing Link

In This Chapter

➤ Frauds and mistakes: Piltdown Man and Nebraska Man

➤ Australopithecus and putting Lucy into perspective

➤ *Homo habilis*—the toolmaker

➤ *Homo erectus*—walking upright and taming fire

➤ Neanderthals—more alike than different?

All the fossil evidence suggests that humans and other mammals are descended from shrew-like creatures that lived more than 150 million years ago. Go back farther, and we find that mammals, birds, reptiles, amphibians, and fish share the same ancestors—humble aquatic worms that lived 600 million years ago. Run the time machine back even farther, and we can trace the ancestry of all plants and animals to bacteria-like microorganisms that originated more than three billion years ago, about one billion years after the Earth was formed. After that … well, no one knows quite where they came from.

Although fossil remains hadn't established earlier precursors for man, Darwin was confident that sooner or later they would turn up. The regions that were most likely to contain the remains connecting man with extinct species, he wrote, "have not as yet been searched by geologists." There was another problem, too—how were

What Does It Mean?

In biology, **fission** refers to a type of asexual reproduction in which an organism divides into two or more parts, each of which develops into a complete organism.

What Does It Mean?

Homo erectus is a hominid that lived about one million to two million years ago and was characterized by its ability to walk upright and make tools.

What Does It Mean?

A **hominid** is a member of the hominidae family of humans and great apes.

anthropologists going to figure out just what a human precursor actually was? The answer wasn't so obvious.

There is no way to test whether today's humans would be capable of interbreeding with those who lived several thousand years ago. As you remember, species are defined as members of a population that can interbreed with one another. (Bacteria and blue-green algae are exceptions to this rule because they do not reproduce sexually, but by *fission*.) If members of a population are no longer able to breed with one another then they become reproductively isolated and are, to all intents and purposes, a different species. Although it's reasonable to assume that people living now could breed with people who lived several generations ago, it's far more difficult to say whether living humans could breed with their ancestors a thousand or a million generations ago.

In fact, ancestors from a half a million years ago—about 20,000 generations—are classified as another species, *Homo erectus*. We are *Homo sapiens*. But when exactly did *Homo erectus* ("walking man") turn into *Homo sapiens* ("knowing man")? The answer isn't clear. Nonetheless, anthropologists believe that it's worthwhile to make the distinction. Genetic changes from one generation to the next are small, but over time those changes add up and if the time span begins to stretch out into hundreds of thousands of years it's safe to assume that you're talking about a different, though related, species. But no one can pinpoint a stage in evolution and say this is where *Homo erectus* stopped and this is where *Homo sapiens* began.

What if we go still farther back—30,000 or 40,000 generations? What came before *Homo erectus?* When *paleoanthropologists* (a profession combining paleontology and anthropology) tried to dig—literally—farther into the past, they came upon more puzzling discoveries. Which hominids belonged among our ancestors, they wondered, and which *hominids* had followed evolutionary paths that turned out to lead nowhere at all? Not only did they have trouble determining the answers, but, even as the fossils mounted up, they even had trouble posing the right questions.

Hoaxes and Frauds

Many of Darwin's contemporaries seized on the lack of a "missing link" as a way of disproving his whole theory of evolution. But for as many nay   sayers, th ably twice as many enthusiasts ready to hunt high and low and find the missing link, certain that fame and fortune would follow. This led to all sorts of mischief. In 1912, a doctor and amateur paleontologist named Charles Dawson discovered an ape-like jawbone and part of a human-looking skull in a gravel pit near Piltdown, England. The skull was very human-like. He assumed that the two specimens were from a distant ancestor called "Dawn Man," estimated to be 500,000 years old. But on further investigation, the jaw turned out to have come from an ape, and the skull, while human, proved to be somewhat younger than Dawson had supposed—500 years old rather than half a million. Piltdown Man was nothing more than an elaborate hoax, but one that, for a brief time, anyway, had a good many people hoodwinked.

Ten years later, Henry Fairfield Osborn, then head of the American Museum of Natural History, declared that a specimen of a tooth that he had been sent from a collector in Nebraska had come from an ancient ancestor of man. A gullible public was eager to buy into supposed evidence of a missing link. The *Illustrated London Daily News* even ran a picture of Nebraska Man, conveniently providing him with an adoring wife. When other parts of the skeleton were dug up in 1927, the my Nebraska Man's true identity was revealed—the tooth had come from an extinct pig.

What Does It Mean?

Paleoanthropologists study paleoanthropology, a combined discipline—anthropology and paleontology—having to do with the study of fossil hominids.

What Should a Proper Hominid Look Like?

In spite of these disappointments, paleontologists relentlessly pursued their search for human precursors. Over time, they unearthed not one, but several fossils of possible intermediate forms between apes and humans Virtually any of the presumed "hominid ancestors" of man that they've found would likely be classified as apes if they walked through the door today. (An ape is defined simply as a tailless monkey.) But how to distinguish evolving apes from evolving humans, or evolving humans from evolutionary dead ends that may have resembled apes was more of a problem. As a result, the effort to find these missing links has been fraught with controversy and bitter disputes as anthropologists and paleontologists have clashed over whether a particular fossil is actually a hominid ancestor of humans.

Paleoanthropologists, of course, needed some criteria to make their call. Several characteristics of these fossil remains are examined to see how closely they resemble humans (or how different they are from us). Much of the anthropological research has focused on the size of the skull, for that reason. Human skulls range in cranial capacity—the amount of brain matter that can fit inside a skull—from 56 cubic inches to almost 160 cubic inches. (The average for women is 90 cubic inches and, for men, is about 101 cubic inches.) Although there is no evidence whatsoever that brain size has anything to do with intelligence, paleontologists nonetheless use cranial capacity as a principal measure of assessing whether the fossil remains might be among our direct ancestors. In addition, scientists look for such traits as the shape of the skull, brow ridges, the location of eye sockets, and the structure of the teeth and jaw—to try to determine whether an ape-like creature is really a close relative of man or a prehuman.

You'd think that, by now, paleontologists and anthropologists would have developed consistent criteria to make this determination. But that is not the case at all. In his book *Lucy,* Donald Johanson had this to say:

> It may seem ridiculous for science to have been talking about humans and pre-humans and protohumans for more than a century without ever nailing down what a human was. We do not have even today, an agreed-on definition of humankind, a clear set of specifications that will enable any anthropologist in the world to say quickly and with confidence this one is a human; that one isn't.

All the same, paleontologists and anthropologists continue to try. Over the years, they have advanced several candidates for the role of our ancestors. Whether we will want to put their pictures in the family photo album is another story.

The names given to the hominids are generally derived from their perceived characteristics or place of origin:

➤ *Pithecus* means "ape"

➤ *Anthro* means "man"

➤ *Homo* means "self" or "modern man"

➤ *Pithecanthropus* means "ape man"

➤ *Australopithecus* means "southern ape"

➤ *Platyops* means "flat-faced"

➤ *Zinjan* means "East African"

What Does It Mean?

Australopithecus is a hominid that lived about four million years ago; its name means "southern ape" because it is ape-like in appearance and was found in South Africa.

Until recently, scientists have recognized only two groups of hominids. The first was believed to be Australopithecus, and the second was us—the genus *Homo* (*habilis* and *erectus*), which originated more than two million years ago and led to modern humans. Some scholars make the case that there was a third group called Paranthropus, which became extinct about one million years ago. (Some anthropologists dispute this classification.) But, as we said, any attempt to classify prehumans or determine where they fit in the evolutionary scale or what their relationship is to *Homo sapiens* almost inevitably gets mired in rancorous debates. "[S]tudents of human evolution, including myself, have been flailing about in the dark," complained Harvard anthropologist Dr. David Pilbeam. "[O]ur data base is too sparse, too slippery, for it to be able to mold our theories .... Paleoanthropology reveals more about how humans view themselves than it does about how humans came about."

Australopithecines: Our Earliest Ancestors ... Maybe

Australopithecus lived four million years ago and had an upright human stance but a cranial capacity of less than 30 cubic inches, which is comparable to that of a gorilla or chimpanzee and just about one-third that of humans. Australopithecus's skull is a curious combination of ape and human characteristics: a low forehead and a long, ape-like face, and teeth with human proportions. It earned its name of "southern ape" because the first fossils were found in 1924 in limestone quarries of the Taunga region of South Africa.

The first fossil was given to Dr. Raymond Dart, a professor of anatomy in Johannesburg, who, on the basis of the shape of the teeth, asserted that it was an early ancestor of man. When his analysis was published, it was greeted by a storm of criticism from scientists who were convinced that it was some kind of chimpanzee. The skull became known as "Dart's baby." Placed side by side with specimens of human and living ape skulls, some critics noted, Dart's baby was so close to an ape as to be indistinguishable from one. Today it is more kindly known as "the Taunga Child."

Whatever it was, Taunga Child was not alone. In the next several years, Dart and others found similar fossils in South Africa. The apes were of two types, suggesting that Australopithecus had followed two lines of development along parallel tracks. One was small and dubbed "gracile" (slender), while the other was larger and so was called "robust." Not everyone was convinced that there were two types; others insisted that they represented only males and females of the same species. Others contended that the gracile form was an earlier ape that had evolved into the robust form. These animals are known as *Australopithecus africanus* and *Australopithecus robustus*.

But that is not the end of the Australopithecus story—or the controversy that its discovery engendered. In most instances, when these fossils were unearthed, they were found along with other animals such as baboons. Many of these other animals had perished violently—many had shattered skulls. Paleontologists assumed that the injuries had come from the bone weapons found near the fossils. The presence of weapons and tools as well as evidence of fire suggested that these ape-like creatures might have been hunters (and killers), except that some of their skulls were bashed in as well. Were they the hunters or the hunted?

213

To some, the answer was clear. In his book *African Genesis*, Robert Ardrey, a journalist, popularized the idea that Australopithecus was the "killer ape." Stanley Kubrick's legendary film, *2001*, only reinforced the idea of the killer ape in the public's mind. But the question of whether Australopithecus was quite so bloodthirsty is far from resolved. Other anthropologists contend that these creatures have gotten a bad rap and that, on the contrary, they were vegetarians—hunted, yes, but not hunters—who used bones to scavenge for food, not to beat in the brains of their prey.

Australopithecines were believed to have been bipedal—they moved on two limbs—and walked upright. But in the 1970s, Richard Leakey, one of the world's leading paleontologists, found several nearly complete remains that called into question their ability to walk upright. That, in turn, threw into doubt their association with humans. Most evolutionists have reached the consensus that both *Australopithecus africanus* and *Australopithecus robustus* were experiments of nature—ancestors of neither humans nor apes—who left no lineages behind them. That would pretty much rule out the possibility that they have a place in the hominid family tree.

The Lucy Show

In 1974, paleoanthropologist David Johanson and French geologist Maurice Taieb discovered half a female skeleton in Hadar, a desert in northeastern Ethiopia. The two estimated their find to be nearly three million years old. They named their fossil Lucy, after the Beatle's tune "Lucy in the Sky With Diamonds." A year later, 13 fossils similar to Lucy were found. They were all called Australopithicines. Because their skulls were markedly more ape-like than either *A. africanus* or *A. robustus,* their discoverers called them *A. afarensis.*

If she stood upright, Lucy would be about $3^1/_2$-feet tall; even by ape standards, she had a tiny brain for a creature of her kind. Johanson had no doubt that Lucy wasn't human—she was too small, her jaw was the wrong shape, and her teeth "pointed away from the human condition and back in the direction of apes." On the other hand, judging from the knee joint and pelvic bones, he was convinced that Lucy was capable of walking in an upright bipedal fashion. That led him to conclude that, in spite of her odd skull and body shape, that she was an ancestor of man as well as of *A. africanus.* His contention was hotly disputed by other anthropologists who maintained that an examination of Lucy's shoulder blade, foot, and hand bones, proved that she must have been a climber and that the proportions of her limbs wouldn't allow for an efficient upright gait. Yet other anthropologists asserted that *A. afarensis* is really the same animal as *A. africanus.*

Lucy Is Given the Shaft

For two decades, *A. afarensis*—Lucy and her assumed relatives—were considered the earliest hominid species. Now it seems that the fossil species *Australopithecus afarensis* had more company on the African plain than anthropologists had expected. Until the extinction of Neanderthals about 28,000 years ago, there were always two or more species of hominid in existence for up to four million years.

In 1999, a research team led by Dr. Maeve G. Leakey found a nearly complete skull and partial jaw while excavating on the western side of Lake Turkana in northern Kenya. Although this skull had distinctly different features from Lucy—it had a flattened face and small molars—it lived in the same era, about three and a half million years ago. Leakey named the new member of the hominid family *Kenyanthropus platyops,* or "flat-faced man of Kenya." (It may not have been a woman; its sex is still to be determined.) Its discovery opened the possibility that there wasn't just a single line of descent stretching through the early stages of human ancestry, but at least two—Lucy and flat-face.

But which of these fossils truly represented early hominids? Was Lucy a progenitor of humans or another evolutionary dead end? Or was she of a different species altogether? At the very least, the existence of Kenyanthropus meant that there were at least two lineages of possible prehumans extending back three and a half million years. That meant that the early stages of human evolution were even more complex than anthropologists and paleontologists had thought.

But then why should humans be any exception to evolution? Many organisms developed along different branches, some of which continued to grow and some of which were cut off by natural selection. When the early hominids split off from ancestors of the ape and started walking on two legs, their increased mobility would have allowed them to migrate into new habitats and ultimately develop into new species. The question that Maeve Leakey's discovery raised was that of these two new species—flat-footed man and Lucy: Which has the right to claim to be our forbearer? One thing is certain: Kenyanthropus is not of the genus Homo, which evolved much later. It seems, rather to be a species all its own.

Who Are They?

The Leakeys are the first family of paleontologists. Louis (1903–1972) and Mary (1913–1996) Leakey, the husband-and-wife team, were pioneers in the search for early hominid fossils in East Africa. Their son Richard (b. 1941) has made more than 200 fossil discoveries; his wife, Maeve Leakey, and daughter, Louise, have both carried on the family's fossil-hunting tradition.

The Handy Man

In 1959, while hunting for hominid fossils in Olduvai Gorge in East Africa, the great paleontologist Mary Leaky—Maeve's mother-in-law—discovered a skull broken into about 100 pieces. She took her find to her husband, Lewis. Lewis, whose paleontological accomplishments rivaled hers, immediately dismissed the skull's significance.

"Why it's nothing but a damn Australopithecine," he reportedly said. Louis changed his mind when he investigated the site of the skull further. Louis made two related discoveries: He found fossils of many other animals, most of which had died violently. The butchery was unmistakable. The bones had been deliberately broken so that the killers could get at their marrow. When he found stone tools at the same site, he decided that the fossil his wife had unearthed was more advanced than he'd initially thought.

What Does It Mean?

A **sagital** is a ridge of bone that runs from the top to the bottom of the skull, found in species of ape or animal-like creatures that may represent human ancestors.

This hominid was both a toolmaker and a butcher, so Leakey dubbed him *Homo habilis,* or "handy man." But *Homo habilis* met with as much skepticism as other contenders for human ancestry. For one thing, *Homo habilis,* who lived about one and a half to two million years ago, had huge molars that didn't look remotely human, and it also had a very small brain and a large bony *sagital* crest on the top of its skull—a ridge of bone running along the top of the skull from front to back. That wasn't a human characteristic, either. Later, Leakey backpedaled, uncertain as to how much of a toolmaker the fossil actually was, and demoted him to the classification of Zinjanthropus, which means "East African Man." Today Zinjanthropus is considered just another robust australopithecine.

But Louis Leakey wasn't ready to give up entirely on *Homo habilis.* In 1964, his research team found four new fossil specimens in Olduvai Gorge. These fossils had larger brains than Australopithicines. So, Leakey said that they deserved to be classified as *Homo habilis,* after all. It was Louis's son, Richard, who finally found the elusive toolmaker that his father had been searching for. In the early 1970s, the younger Leakey discovered several fossilized bone fragments of a skull, which was given the unpoetic name KNMER 1470—its registration number at the Kenya National Museum. This skull, with a capacity of about 45 cubic inches, was considerably larger than the skulls of most ape-like creatures previously found. There were only small eyebrow ridges, no sagital ridge, and a domed skull typical of a human.

In addition, Leakey found several bones belonging to 1470 that also were remarkably human in appearance. That 1470 was almost three million years old posed a problem for Lucy. Johanson, who had discovered Lucy, had maintained that *A. Afarensis* was the sole evolutionary link between apes and man. But the discoveries of the Leakeys—Louis and Maeve two decades apart—both threatened to topple Lucy from her perch and throw open the field of contenders for our earliest ancestors.

Homo Erectus and the Cost of a Larger Brain

No sooner had Darwin published *The Origin of Species* than a Dutch physician named Eugene Dubois set out to find the "missing link" between apes and man. In 1891,

Dubois met with success. He discovered a collection of fossils—specifically, two large molar teeth, an ape-like skullcap, and a human femur—in Java in the Indonesian archipelago. He concluded that they all must belong to the same creature that he described as "admirably suited to the role of the missing link." He called this specimen *Pithecanthropus erectus* ("upright ape-man"), which, without the benefit of precise radiocarbon dating, he declared was about 10 million years old. (They are, in fact, much younger than that.) *Homo erectus* is also known as Java Man in recognition of the location where the fossils were originally found.

G. K. Chesterton, the noted English essayist, remarked on the publicity that *Pithecanthropus erectus* received with a certain astonishment:

> People talked of Pithecanthropus as of Pitt or Fox or Napoleon. Popular histories published portraits of him like the portraits of Charles I or George IV .... No uninformed person, looking at its carefully lined face, would imagine for a moment that this was the portrait of a thigh bone, a few teeth, and a fragment of a cranium.

Fossils of *Homo erectus* have subsequently turned up in a variety of locations—East Africa, South Africa, Ethiopia, and Java. The oldest specimen comes from Lake Turkana, a veritable treasure trove of hominid fossils, and is almost two million years old.

Where Has Nellie Gone?

In the annals of hominid finds, an example of *Homo erectus* called Peking Man occupies a unique place—or, at least, it did until the fossils mysteriously vanished. In 1929, an almost complete hominid skullcap was found in a limestone cave in Choukoutien near Peking, China. The ape-like skullcap was similar to the one found in Java by Dubois. By the beginning of World War II, paleontologists had excavated fragments of 14 skulls, 12 lower jaws, and 147 teeth.

Many of these bone fragments were collected from scattered locations and were assembled to form a skull. The skull and face bones, for example, came from a level 85 feet below the jawbone attached to it. Then a sculptor was called in to superimpose a woman's features on a cast of the skull. The model that resulted was affectionately nicknamed "Nellie." Then, as the Japanese were closing in on the capital (now known as Beijing), a U.S. Marine stationed in the city placed Nellie in his footlocker, evidently with the intention of shipping the fossils to America. They were never seen again, and their fate has been a source of speculation ever since.

More recent excavations carried out by the Chinese in Choukoutien have turned up more than a thousand fragments of stone tools, the skulls of more than 100 animals, and fragments of six *Homo erectus* skulls, apparently smashed in by clubs. Here, too, scientists found evidence of fire-making technology. The Chinese scientists could only conclude that the cave dwellers were *Homo erectus*. *Homo sapiens* did not exist 500,000 to a million years ago.

The Mystery of Turkana Boy

Then in 1984, Richard Leakey made yet another astonishing discovery when he came on a nearly complete fossilized skeleton of an obviously human 12-year-old boy in the fossil-rich region of Lake Turkana in Kenya. Except for certain details of the skull, the skeleton resembled a modern human in practically all respects. Leakey suggested that the boy would have gone unnoticed in a crowd today. What made the find so remarkable was that Turkana Boy, as he became known, was found in strata about 1.6 million years old. In spite of its human characteristics, Turkana Boy is classified as *Homo erectus*.

The discovery of Turkana Boy allowed scientists, who previously had only bone fragments to work with, to put together a more detailed portrait of *Homo erectus*. *Homo erectus* had human-like features and a larger body build not unlike our own, and he was endowed with a large brain—his skull was twice the size of Lucy's, giving him the intelligence to make sophisticated tools. Judging by Turkana Boy's bone structure, with its thin arms and legs, anthropologists believe that *Homo erectus* would have been admirably equipped for surviving on the African plane. Unlike *Homo habilis*, *Homo erectus* was nimble and quick, able to get to a carcass before animals could devour it. As a toolmaker, he was especially proficient. He'd learned the secret of making tools that had razor-sharp edges—a technological breakthrough—that greatly expanded the range of tasks that he could do. At the same time, *Homo erectus* had developed an ability to exploit fire that also would have given him a great advantage over animals. By taming fire, *Homo erectus* could see in the dark, cook food for the first time, and scare away predators.

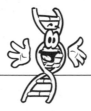

A Natural Selection

Lake Turkana, located in the Great Rift Valley of eastern Africa, has been the site of many significant hominid fossil finds, beginning in 1967.

Nothing in nature is cost-free, though; *Homo erectus* had to pay a price for his larger brain. The brain could not be fully developed at birth—that would have made it impossible for the fetus to get through a pelvis that was only slightly larger than the diminutive Lucy's. So, the brain had to develop outside the womb—as human babies' brains do—expanding to three times its size at birth. Newborn baboons and other animals, by contrast, have almost fully formed brains. That allows the baboons to be very active and independent shortly after they are born.

Babies born to *Homo erectus* were completely dependant. This means that *Homo erectus* had to come up with an evolutionary solution, which humans would make use of. That solution was some kind of family organization to look after the offspring. *Homo erectus,* of course, still had to scavenge for food as well as raise children. (Modern couples struggling to balance the demands of work, home, and children might sympathize.) *Homo erectus* seemed to have negotiated the conflicting needs quite well. In fact, in evolutionary terms, *Homo erectus* is really something of a success story because they remained essentially unchanged for nearly a million years—at least, until the appearance of a hominid species that looked and acted very much like us.

The Neanderthals Tell Their Side of the Story

Neanderthals are closest to us in appearance and in time than any other hominid. They lived as recently as 24,000 years ago, making them contemporaries of humans, although their lineage extends back about 200,000 years. (Human beings emerged about 50,000 years ago.) Although first identified in 1856 in Germany, Neanderthal Man had actually revealed himself earlier in the form of fossil remains in Belgium in 1829 and on Gibraltar in 1848. It was just that, until the German discoveries, no one had realized exactly what the earlier fossils represented. Subsequently, fragments of some 60 Neanderthal skeletons were found in locations as far away as China, central and north Africa, Iraq, Czechoslovakia, Hungary, Greece, and northwestern Europe.

Neanderthals Get a Bum Rap

For many years afterward, *Neanderthals* were commonly depicted as dim-witted brutes, based on early—and flawed—reconstructions of the fossils. Typically, a Neanderthal differed from humans in that he had a heavier build, a low forehead, a large nose, a recessive chin, and a prominent brow ridge with a bony arch protruding over each eye.

Later investigations demonstrated that Neanderthals were fully upright and walked the same way that *Homo sapiens* do. If anything, their cranial capacity was even somewhat larger than humans can boast of. (The difference is accounted to the Neanderthal's greater muscular mass.) From the evidence found at the fossil sites, Neanderthals exhibited a much higher level of sophistication than any of the earlier hominids had. They appear to have commemorated the dead with symbolic rituals, for instance. They also produced ingenious stone tools, using primitive casting models, capable of making several types of implements.

Neanderthal Man was the first "ape-man" to have been discovered in Darwin's day. When the first finds turned up in Germany, little attention was paid to them. But that all changed with the publication of Darwin's *Origin*. Suddenly the search was on for the so-called missing link. Naturally, a controversy broke out as to whether these Neanderthals represented the link. Darwinians argued that Neanderthals were ape-like progenitors of the human species, while anti-Darwinians countered that these individuals were fully human. If there

What Does It Mean?

The name **Neanderthal** is derived from the location where fossils of Neanderthals were first found in 1856, in the Neander Valley in Germany. *Tal* means "valley" in Germany.

were some differences in their form, the critics said, it must have been due to a disease such as rickets or arthritis.

The Neanderthals Get a Refurbished Image

Later in 1908, when more Neanderthal skeletons came to light in the French villages of LeMoustier and La Chapelle-aux-Saints, the debate took a new direction. More extensive examination of the remains convinced some anthropologists that Neanderthals were an inferior species closer to apes than to any other human group. This demeaning view of Neanderthals prevailed until the mid-1950s. It was then that anatomists went back and took another look at the remains excavated from La Chapelle-aux-Saints. Lo and behold, anatomists declared, it appeared that the fossils of Neanderthals that they examined had suffered from severe arthritis, which had affected the vertebrae and caused them to slump—anatomical details that had helped form the earlier erroneous conception of Neanderthals as being unable to walk upright.

The anatomists offered a completely different image of Neanderthals, minimizing the differences between them and us: "If he could be reincarnated and placed in a New York subway, provided he were bathed, shaved, and dressed in modern clothing, it is doubtful whether he would attract any more attention than some of its other denizens." The revisionist view is now the accepted one. Today Neanderthal Man is classified as *Homo sapiens,* putting him in the same species with us. Nonetheless, the jury still seems out on how to classify them. Some anthropologists feel that its advocates have gotten carried away in their defense and that Neanderthal is too primitive to be granted the privilege of belonging to *Homo sapiens.* "[T]he situation remains fluid," is how *Smithsonian Magazine* summed up the debate so far.

Here is the hominid cast of characters:

➤ Australopithecus (three million to four million years old)

    A. africanus

    A. robustus (also Paranthropus)

    A. afarensis (Lucy)

    Taung Child ("Dart's baby")

➤ *Kenyanthropus platyops* (three million to four million years old)

➤ *Homo habilis* (two million years old)

➤ *Homo erectus* (500,000 to 1 million years old)

    Peking Man

    Turkana Child

    (also *Pithecanthropus erectus* or Java Man)

➤ Neanderthal Man (24,000–200,000 years old)

➤ *Homo sapiens*—modern man (50,000 years old)

The Least You Need to Know

➤ Because the fossil record of Darwin's time contained almost no examples of possible human precursors, many critics of evolutionary theory believed that such a "missing link" did not exist.

➤ Since Darwin's time, paleontologists have succeeded in unearthing many fossil remains, some fragmentary and some more complete, that date back as far as four million years.

➤ Until recently, the oldest ancestor was believed to have been Lucy, the skeletal remains of a diminutive girl with a small brain who lived about four million years ago.

➤ About two million years ago, a hominid dubbed *Homo habilis*—"handy man"— emerged in eastern Africa.

➤ *Homo erectus* could walk upright, make razor-sharp tools, and tame fire, but its larger brain size meant that babies were born dependent—unlike baboons, for example—so that their brains and bodies would have a chance to develop outside the womb.

➤ Neanderthals lived as recently as 24,000 years ago, making them contemporaries—for a while, anyway—with *Homo sapiens*.

I DON'T WANNA GO!

Chapter 20

Climbing Down from the Trees

In This Chapter

➤ How *Homo sapiens* originated in Africa

➤ Early hominid migrations from Africa

➤ *H. sapiens* and Neanderthals—why did one win and the other die out?

➤ How tool making spurred evolution

➤ Why human evolution has stopped—and will it resume again?

With all the controversies raging over which prehuman or protohuman has the right to claim to be our ancestor, you can begin to see why tracing our roots back so far into the past poses a daunting challenge to anthropologists and evolutionary biologists. Unfortunately, as we've already seen in the case of the Neanderthals, things don't become much clearer even when scientists try to pinpoint how and when human sapiens—our species—began to emerge in history. Determining the roots of anatomically modern humans has long been a puzzle to students of human evolution.

One of the major stumbling blocks is an absence of fossil evidence. So far, no human skeletal remains have been found that date back more than half a million years. Those remains from that period have been classified not as *Homo erectus*, our immediate predecessor, but rather as a form of *Homo sapiens* classified as "archaic" to distinguish them from anatomically modern humans. Most anthropologists believe that these early humans came from Africa, which would account for why paleontologists have made as many hominid discoveries on the continent, especially in the area in the Great Rift Valley of Kenya. Based on DNA evidence, biologists have placed the origin of all modern humans somewhere between 140,000 and 290,000 years ago. Studies of a particular part of the human chromosome involved in the immune response suggest that the early population consisted of a paltry 500 to 10,000 individuals. We owe our existence to them.

So how did such a small group of individuals manage to overwhelm more populous species? Why did humans evolve and leave so many other species of *Homo genus* and Neanderthals in the dust? And why, for that matter, having evolved to the extent that we have, are we no longer—at least, temporarily—on the march again? While environmental factors are causing evolutionary changes to individuals, and even to whole populations, significant human evolution on a large scale hasn't occurred for upward of 250,000 years. But will that remain the case indefinitely? That is a question that scientists from many different fields are struggling to come to terms with today. Nothing more or less is at stake than our ultimate fate.

Out of Africa: The First Migrations

Once early humans left Africa, where did they go? Archaic human remains outside Africa are scattered all over the place—a skull belonging to an archaic *Homo sapiens* was found in China dating back 200,000 years; skulls of similar age have been unearthed in France and in Europe. Others about 100,000 years old have been found in the Middle East. Obviously, at some point, humans—or prehumans—left their ancestral home in Africa and began the long trek north, west, and south.

One hypothesis suggests that most of the original hominid stock migrated from Africa at two different periods. One lineage, about 700,000 years ago, became the ancestors of the doomed Neanderthals who relocated to the temperate zones of Europe and the Middle East, where they continued to evolve. According to this hypothesis, a second lineage, which evolved in Africa about 100,000 years ago, eventually evolved into modern humans. Migrants from Africa colonized China about 68,000 years ago, Australia at least 50,000 years ago, and Europe 36,000 years ago. Modern genetic studies, published as *The History and Geography of Human Genes,* classifies humanity into four major ethnic regions: African (Khoisan), Caucasoids (Basque), Mongoloids (American Indian), and Australians (Aborigine).

What motivated these migrations is a subject of speculation and debate. Why did some species end up in China and others go to Europe or the Middle East? No one knows. Making the situation more complicated is that so many different types of prehumans, early humans, and Neanderthals were spreading over the globe virtually simultaneously. For example, early modern humans (Cro-Magnon Man) who settled in western Asia about 100,000 years ago were both much older and genetically distinct from the Neanderthals who also migrated there. But they were much older than the Cro-Magnon peoples who spread throughout Europe 30,000–35,000 years ago as well.

Biochemically, Asian and European populations appear to be more similar to each other than those of either group are to African populations. That leads evolutionary biologists to believe that Asians and Europeans are more likely to have shared a common ancestry some 40,000 years ago, but their common ancestry with African populations extends back almost three times as long. That reinforces biologists' conviction that all modern *Homo sapiens* owe their origin to Africa. Even so, as the genetic evidence makes very clear, the genetic variation among modern human populations is small compared, for example, with the variation between apes and monkeys.

What is indisputable is that environmental change played a significant part in influencing the dispersal of early humans (and others in the hominid family) and probably had a role in the way in which they evolved—or died out. Around 186,000 years ago, for instance, while many of these great migrations were going on, an Ice Age set in, creating arid conditions in Africa. Ice extended from present-day Great Britain eastward to Scandinavia and Siberia; other glacial systems advanced over the Alps and the Himalayas as well as the Andes. At the peak of the Ice Age, Northern Germany, England, and Ireland lay submerged under almost 4,000 feet of ice. All these glaciations effectively choked off further migration from Africa until they began to recede about 120,000 years ago.

The first modern humans are believed to have migrated into Europe about 40,000 years ago. They came with the expertise derived from two different tool traditions. South of the Sahara, archaic humans still relied on Stone Age tools developed between 200,000 and 40,000 years ago. But in North Africa, a new tool tradition had taken hold—also about 40,000 years ago—associated with early *Cro-Magnon* cultures. At the time, the climate in North Africa was quite different from what it's like today: Even the Sahara desert, now sparsely inhabited, was a lush region, filled with lakes, streams, vegetation, and plenty of game. This hospitable environment spurred technological development. Hunters produced better weapons—pointed spears, bows and arrows, and wooden sickles with flint barbs. Archaeologists have found that these people also used grinding stones to grind down seeds, evidence of early agricultural experimentation.

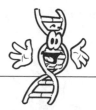

A Natural Selection

During the four-billion-year history of the Earth, it has been subjected to several ice ages, each lasting tens of millions of years. At one time or another, ice has covered from 10 to 50 percent of the planet's surface.

What Does It Mean?

Cro-Magnon is the name given to early *Homo sapiens* who inhabited western and southern Europe during the last ice age about 120,000 years ago. They are distinguished by high foreheads and well-defined chins.

Why Did the Neanderthals Vanish?

Modern humans and Neanderthals appear to have come into contact with each other between 100,000 and 40,000 years ago. How the two got along is unknown. (The best-selling novel *Clan of the Cave Bears* offers a rather fanciful scenario of what might have happened.) Neanderthals and modern humans, while composing distinct populations for at least 60,000 or 70,000 years or longer, still had a great deal in common. They both shared three million to four million years of evolutionary history, and they had comparable brain size and structure. And, like humans, Neanderthals had an ability to make complex tools, providing additional evidence that they had a fairly high intellectual capacity. There is a good deal of debate about whether Neanderthals and modern humans were genetically similar—and thus able to interbreed—or whether the Neanderthals were reproductively isolated from humans. (This is one of the reasons why anthropologists are still divided about how to classify Neanderthals—are they *Homo sapiens,* or do they belong to another lineage?)

A Natural Selection

The Neanderthals stained their dead with red ocher, which suggests to archaeologists that they might have had some belief in the afterlife.

Whether they were genetically distinct from humans or not, anthropologists believe that Neanderthals did not breed as quickly as humans did. As a result, humans soon outnumbered them in Europe, overtaking them about 40,000 years ago. Some recent archaeological finds suggest that the Neanderthals weren't quick enough on their feet—literally. For some reason, about 35,000 years ago, they failed to undertake seasonal migrations in search of food.

Migration, it turns out, is one of the pivotal forces that can enhance evolution and, consequently, heighten a population's chances of survival. A migratory lifestyle places demands on a population, forcing it to improvise and develop efficient modes of transportation. That, in turn, requires the capacity to plan ahead—what do we do if we run into a storm or if the game is less plentiful than last year? It also is a means of transmitting information and maintaining social organization. A population that was ready to look far and wide for sources of food, whatever the temporary inconvenience, would have had a distinct advantage over a domesticated population entirely dependant on what they could get at home. The widening availability of food then would have triggered population explosions and brought about such social changes as the invention of art, jewelry, and more sophisticated tools.

So even if Neanderthals were as intelligent as humans, some scientists say, they somehow neglected to exploit the opportunities that humans found by periodically setting out from home. As time went on, Neanderthals lost their edge, becoming more technologically and culturally isolated. Unable to maintain a birth rate as high as the early humans, they would have found themselves pushed into an increasingly tight social and physical corner. Why did the Neanderthals stick so close to home and condemn themselves to extinction? No one can say. What the archaeological evidence suggests is that humans didn't triumph at the expense of Neanderthals because humans were biologically superior, but rather because humans, unlike Neanderthals, were willing to pursue an adaptive strategy that gave them an overwhelming technological and social advantage.

Gearing Up for the Stone Age: Bring On the Tools

Adaptation, after all, was what caused early hominids to diverge from apes. Which species of hominid might have produced the first stone tools is unclear, although some scientists believe that *A. robustus* is a leading contender for the honor. Other scientists say that it was an early form of the *Homo genus,* pointing out that smaller-toothed humans would have needed tools more than *A. robustus* because they had to chop up their food to chew it. *A. robustus,* with his larger molars, wouldn't have needed to trouble with such niceties.

In addition, a tool-making human would have had an advantage during a period marked by a cooler and drier climate that occurred about 2.8 million to 2.4 million years ago. As vegetation became sparser, as a result of the climactic change, finding alternative food sources such as roots and tubers would have become more crucial. But, of course, it was necessary to get them out of the ground first, which is where tools would have come in handy. Hunters, too, would have needed tools to improve their prospects of catching game. Other scientists dispute this theory, pointing out that the climate in this period underwent several shifts: It didn't just become drier; at times it became wetter. In that case, early humans would have had to adapt to a variety of unpredictable climates.

Whatever the origin of tool-making skills, they seemed to have propelled the evolution of early humans in several ways, not the least an increase in brain size and the capacity to communicate through language. Although early tools were simple stone implements, over time they became very complex. As tools became more complex—beginning with the first sharp-edged stone tools of early hominid—they could be constructed only using several different components as well as a variety of materials such as bone, flint, and wood. And the more complex these tools, became the more they required the contributions of several individuals to produce them.

To ensure coordination in the process, some form of sophisticated communication needed to take place—which is how some scholars believe language may have gotten started. (How language developed—whether it is based on an innate "hard-wired" capacity in our brains or is acquired through learning alone or by some combination—is a fascinating and controversial subject that we haven't the time or space to go into here.) By the end of the Paleolithic Age, early humans were producing spear throwers, harpoons, heat processes to manipulate raw materials, and ceramic vessels.

The Stone Age is divided into three periods: Paleolithic, Mesolithic, and Neolithic:

➤ The Paleolithic Period (Old Stone Age) began about two million years ago, with the development of stone tools by the first prehumans, and ended about 13,000 years ago at the end of the Ice Age. Simple tools such as flaked stone implements and chipped pebbles, were used for agriculture and hunting.

➤ The Mesolithic Period (Middle Stone Age) began about 13,000 years ago. As the Ice Age receded and the climate warmed, early humans were freer to migrate farther into temperate and tropical forests, where more resources were available for tool making.

➤ The Neolithic Period (New Stone Age) began about 10,000 years ago (8000 B.C.E.) with the spread of agricultural villages through the Middle East and Central and South America.

Humans Aren't the Only Tool Makers

Are humans distinguished from primates by their ability to make tools and communicate symbolically? Some recent research calls that notion into question. Wild chimpanzees, for instance, have been observed making and using primitive tools. Animals are even able to use tools to make other tools. Chimpanzees, orangutans, bonobos (a type of monkey), and gorillas have all learned how to use sign language, meaning that they have at least some capacity to communicate symbolically.

Whether these primates can (or would wish to) successfully teach their offspring how to use sign language is being investigated. Even self-recognition, which used to be thought a phenomenon unique to humans, has been shown in various animals. Dolphins, for instance, can look in mirrors and realize that they are looking at themselves, not another dolphin. Where humans do seem to enjoy superiority, however, is in their unexcelled ability to use language.

Why Did Humans Evolve?

Some extraordinary development must have occurred in the distant past to cause Australopitheous to go one way and leave the ape to develop on its own. Not unexpectedly, scientists disagree on the cause, but virtually all of them are convinced that the environment must have had a good deal to with it. The key, scientists believe, is understanding how bipedalism came about. Solve that question, and the rest may fall into place. How did early hominids leave their perches in the trees and, instead of scrambling about on all fours, evolve into beings that walked upright? Three principal hypotheses have been advanced:

➤ The savanna hypothesis

➤ The woodland-mosaic hypothesis

➤ The variability hypothesis

The Savanna Hypothesis

The savanna hypothesis maintains that somewhere between eight million and five million years ago, the climate cooled and became drier. This climate shift caused the African forests to shrink in the east while leaving the forests in the west relatively unscathed. This had the effect of forcing the east African ape population to adapt, while leaving their brethren in central and west Africa to carry on as before. With fewer trees to escape up into, the east African apes had to figure out how to acclimate to open *savannas*. Solitude might have worked well in the treetops, but not in the grassy terrain that they now faced. So, social organization became necessary, but it could be achieved only by the development of effective modes of communication. To find prey and avoid becoming prey themselves, these apes learned to walk on two feet. And as they continued to cope with new challenges, their brain size began to increase.

That all sounds good, but paleontologists have discovered that the differences in forest density wasn't very different between eastern and western Africa during the time in question. And recent research also indicates that savannas were not very common in Africa until less than two million years ago, long after the separation between hominids and apes occurred. So, the savanna hypothesis was suddenly on, well, very shaky ground.

What Does It Mean?

Savannas are grasslands found in the tropics, with a scattering of shrubs as well as small and large trees. Savannas are formed by periodic fires, caused by lightning or humans, or by warming climates.

The Woodland-Mosaic Hypothesis

Dissatisfaction with the savanna hypothesis led to an alternative: the woodland-mosaic hypothesis, which proposes that Australopiths evolved in terrain that was partly woodland and partly grassland—a mosaic. This terrain would have given them the opportunity to find food both in trees and on the ground. Having to maneuver between these two habitats would promote *bipedalism*.

The Variability Hypothesis

By contrast, the variability hypothesis states that Australopiths had to contend with a wide range of environments—forests, open-canopy woodlands, and savannas. As a result, they would have had to adapt to several different habitats, leading to the evolution of versatility through adaptation, including bipedalism.

What Does It Mean?

Bipedalism means walking upright on two feet. **Quadripedalism** means moving about on four limbs.

"Walk on Your Own Two Feet!"

Why was bipedalism such an advantage? Experts have suggested any number of answers to the question (many of them probably wrong):

> ➤ It freed the hands, making it easier to carry food and tools.

> ➤ It allowed early hominids to see over tall grass and let them know if a predator was nearby.

> ➤ It made it easier to hunt and reduced their exposure to sun.

None of these ideas has proven especially attractive to scientists, who are more intrigued by studies of chimps that may shed more light on how hominids came to develop bipedalism. When they feed on leaves or forage for fruit, chimps rely only on their legs, suggesting that Australopiths might have done the same. However, as soon as feeding is done, chimps quickly revert to their more characteristic four-limbed—quadrupedal—movement. Early hominids, however, continued to walk upright. Whatever the explanation for bipedalism is, it allowed Australopiths to cover vast distances quickly—something apes cannot do—and yet preserved their ability to climb trees to escape predators when necessary. That Australopiths typically had long, powerful arms and curved fingers suggests that they were good climbers, just as the shape of their pelvis and lower limbs testifies to their bipedalism.

Early human evolution was marked by several important developments:

> ➤ Bipedalism

> ➤ Adaptation to different habitats and climates

> ➤ Tool making

> ➤ Fire making and cooking

> ➤ Seasonal migration

> ➤ Social organization

> ➤ Symbolic communication through language

Why Did Human Evolution Stop?

Phyletic evolution is a term that refers to cumulative and important changes in the gene pool of a population that can lead eventually to speciation—the creation of new species—and the higher levels of change on a macroevolutionary level. The fossil record is actually a record of phyletic evolution. It is difficult to identify any unequivocal evidence of phyletic evolution in humans for a very long time. To find the last time that the fossil record shows any significant change in human phyletic evolution, you'd have to go back to the late Middle Pleistocene Era about 250,000 years ago. That last change occurred during the transitional period between the disappearance of *H. erectus* and the ascendancy of the first humans. Unfortunately, we know almost nothing about this transitional period because of the incompleteness of the fossil record. It is clear, though, that early humans didn't arise fully formed from their *H. erectus* ancestors.

So why aren't humans showing evidence of evolutionary change today? Well, for one thing, humans are a very adaptable lot. No matter how harsh the climate, how fallow the soil, or how remote the location, humans have somehow managed to survive in it. Most species inhabit a particular ecological niche—polar bears would wilt in the tropics, and tuna would die on land. Humans, however, are capable of putting down roots anywhere, even adapting to ocean depths or the reaches of outer space. The capacity to adapt to such an expansive ecological niche required considerable flexibility to develop in the human gene pool. In this case, the saying "the whole world's your oyster" happens to be accurate. So, unless a change in the gene pool was truly momentous, it would probably go unnoticed in a gene pool that already contains such enormous diversity.

What Does It Mean?

Phyletic evolution refers to cumulative and significant changes in a gene pool that can lead to the creation of new species.

There is another reason why humans haven't undergone perceptible evolutionary changes in a long time. That's because humans have developed a culture that forms a barrier between humans and their environment. When we talk about "culture," we're referring to a number of factors that range from technology and science to agriculture and art. Culture initially took the form of crude tools and possibly primitive skin clothing. Now we have pesticides, antibiotics, air-conditioning, and modern medical therapies, to name just a sampling, that prevent environmental influences from having the same impact on humans as they would otherwise. Other species, lacking a sophisticated culture, simply do not have the same capacity to block or mitigate the effects of environmental change.

There's a third reason why evolutionary change appears to have been arrested in humans: the size of the gene pool itself and the relatively extended length of a human generation. In small populations, as we noted earlier, evolutionary changes can occur more rapidly than in large populations. The smaller the gene pool, the fewer genes will need to be changed to modify a population. By the same token, evolutionary change tends to take place in populations that mature and reproduce quickly—fruit flies, for instance, or mice. But humans take several years to mature and develop the capacity to reproduce. That would seem to suggest that the flow of beneficial genotypes, however advantageous, must be extremely slow in humans.

Why Human Evolution May Get Going Again

Phenetic evolution refers to the changes that occur in response to environmental factors, but it is the consistency of these stimuli through time that determines whether the changes will become permanently impressed on a group's genetic inheritance (its phyletic record). In other words, if the changes don't endure because of continual stimulation from the outside world, they won't stick around for very long in the gene pool. It's the difference between being exposed to a chest x-ray or a nuclear meltdown. One will have a transient and most likely harmless effect; the other can cause serious damage or fatalities. And because it does happen on a smaller scale, phenetic evolution is more subtle than phylectic evolution. Although some environmental changes may be observable on living populations, most of them don't persist long enough to show up in the fossil record. Phenetic evolution is the mechanism responsible for the formation of subspecies, races, or varieties.

What Does It Mean?

Phenetic evolution refers to evolutionary changes that occur in response to environmental factors. It is primarily observed on a smaller level in individuals and various groups.

Even though human phyletic evolution might have come to a stop—at least, for the time being—we can't state definitively that the environment will never affect human development again or that further phyletic evolution won't occur. That's because phenetic evolution is still going on. We know that because it can be observed. It's possible that, over time, the accumulation of these small changes will lead to much larger changes in the whole human gene pool and thus spur a resumption of evolution on a larger scale.

We can see phenetic change at work in the way in which diseases spread or increase or decrease in virulence. Some environments appear to encourage disease-producing organisms—the pathogens; other environments are prone to inducing changes in the organism that they infect—the host—and make them more or less susceptible. Several diseases, such as the plague, tuberculosis, and scarlet fever, are influenced by changes in the environment as well as by genetic changes in the host organisms. Some populations, because of repeated exposures to these diseases, may escape their ravages completely, or suffer relatively little distress; others without natural immunity may be virtually wiped out.

For instance, many Polynesian peoples lack a natural immunity to measles and were almost destroyed by measles, to which they hadn't been previously exposed. Highly lethal worldwide epidemics, such as AIDS, the 1918 flu pandemic, or the Black Plague of the Middle Ages, could act as potent agents of natural selection by radically reducing the human population and leaving only a comparatively small number of survivors with natural immunity. In that case, the cultural barriers that we've thrown up to keep potentially destructive environmental forces at bay might prove unequal to the task, setting in motion a new wave of phyletic evolution.

The effects of disease are only one example of how further evolution in humans may occur. Experiments with laboratory animals suggest further, and not necessarily welcome, possibilities. Scientists have shown, for example, that they can induce several kinds of pathologies in lab animals—both social and physical—simply by cooping up too many mice in a cage for a long period of time. Crowding, they found, can cause infertility, cannibalism, mental aberration, and premature death.

Pollution can bring on its own share of ills in lab animals as well. That doesn't mean that when humans are exposed to the same environmental influences, as they often are, they will suffer the same effects as the animals do. Nor can anyone be certain that these pathologies, if distributed widely enough among people, will promote human evolution to any significant degree. (Remember that we said that evolution doesn't move in any particular direction, nor is it to be confused with the idea of progress. So, in that sense, even a destructive influence, such as the plague, can act as an agent of evolutionary change.)

There is yet another possible cause of future evolutionary change—a subject that we'll explore in our final chapter—and that is the actions of humans themselves.

The Least You Need to Know

➤ Based on fossil and genetic evidence, most scientists agree that humans originated in Africa and started to spread throughout Europe, Asia, and Australia in a succession of migratory waves, beginning about 700,000 years ago.

➤ Several types of hominids and early or archaic humans (*Homo genus*), including *Homo erectus* as well as Neanderthals, were living at the same time.

➤ *Homo sapiens* evolved to a greater degree than competitive species because they were better able to adapt to a wide range of habitats, climates, and changing conditions.

➤ The evolution of bipedalism (the ability to walk upright) allowed humans to cover great distances quickly, as well as climb trees to escape predators.

➤ Migration, farming, and hunting as well as the development of tools fostered the need for more social cohesion and communication, spurring the development of language.

➤ There are two basic types of evolutionary change—phyletic and phenetic: Phyletic change occurs on a large scale to a species, while phenetic change occurs on a smaller scale in individuals or populations as a result of environmental influences.

Part 5

The Impact of Evolution on Society

The theory of evolution has stirred more than its share of controversy since its introduction at a meeting of British scientists in 1859. Evolution challenged deeply held religious beliefs. How, critics asked, could man have descended from apes? That wasn't what Darwin's theory said at all (humans and apes branched off from a common ancestor several million years ago), but the notion captured the public imagination and the damage was done.

But, if anything, Darwin's theory has suffered even greater injury at the hands of some of its most ardent supporters who distorted the idea of "survival of the fittest" to justify war, colonization, and the subjugation of the poor by the wealthy.

Although virtually all scientists are in agreement that evolutionary theory provides an accurate description of how nature works, acceptance of evolution has been much slower in coming among the general population. In Part 5, we explore the views of creationists and proponents of a theory called intelligent design and find out why, and on what grounds, they either reject, or question Darwin's ideas.

Finally, to conclude our story, we will engage in a bit of speculation and try to see where evolution is headed. Evidence is growing, for instance, that humans are taking an ever-greater role in shaping evolutionary change by genetic tinkering, which puts us in the unusual situation (or is it a predicament?) of being both beneficiaries and instruments of natural selection.

The Siren Call of Social Darwinism

When *The Origin of Species* was first published in 1859, it proved to be an irresistible source of ammunition for many political and social thinkers of the day. Here, at last, was a theory that could be used or misused to justify political and social agendas. If natural selection applied to other species, why shouldn't it apply equally well to humans? If that were the case, then—or, so the thinking went—it should be possible to derive "natural laws" to account for various inequalities in society. Didn't natural selection pick the "fit" and weed out the "unfit," based on whether they had desirable or undesirable traits? That this view reflected a complete misinterpretation of Darwin's ideas was hardly relevant because many of the nineteenth century political philosophers were interested only in what Darwin could do for them, not what they could do to explain Darwin to the public.

First, we should emphasize that Darwin had nothing to do with social Darwinism, in spite of being linked by name to it. He himself seemed startled at the uses to which his theory had been put. In a letter written to an advocate of social Darwinism 10 years after the publication of *Origin*, he said:

> You will really believe how much interested I am in observing that you apply to moral and social questions analogous views to those which I have used in regard to the modification of species. It did not occur to me formerly that my views could be extended to such widely different and most important subjects.

Who Are They?

Jack London (1876–1916) was a popular American novelist whose vivid and realistic stories of adventure brought him millions of devoted readers all over the world. In addition to *The Call of the Wild,* he gained fame for *The Sea Wolf,* based on his adventures as a sailor on a hunting ship, and *Martin Eden,* an autobiographical novel.

Who Are They?

William Bagehot (1826–1877) was a British economist and political thinker and editor of the *Economist.* He is considered one of the most influential journalists of his day.

The Survival of the Fittest Revisited

Simply defined, social Darwinism is a nineteenth century theory that proposes that individuals, groups, and races are subject to natural selection. The theory was based on the idea that a struggle in nature was going on continuously and that only the "fittest" would survive. (We've seen that Darwin's idea of the "fittest" referred to reproductive success, not simply to the capacity to triumph over a competitor or survive in the wild.) According to social Darwinists, if certain people were weak or fell behind in social or economic status, that must mean that they were somehow unfit and that nature was weeding them out of the population.

The theory of social Darwinism seemed to work equally well when applied to the political struggles that were taking place in Europe and elsewhere at the time. Like human species, nations were locked in a struggle for survival. Civilized nations—those that had inherited beneficial traits from their ancestors—had an obligation to supplant barbarous nations. Readers of social Darwinist tracts had no difficulty understanding that "advanced civilizations" were white and of European descent.

Social Darwinism caught on very quickly with the public. In 1903, the popular novelist (and passionate social Darwinian), Jack London, published *The Call of the Wild,* to great acclaim. The book tells the story of how a domesticated dog named Buck—a cross between a Saint Bernard and a Border collie—is stolen and taken to the Alaskan Klondike. Forced to survive in the frozen wastes of the north, the dog learns the hard (social Darwinian) lesson: "Kill or be killed, eat or be eaten was the law and this mandate, down to the depths of Time, he obeyed." After many bloody adventures, Buck not only survives but even becomes the leader of the pack of purebred wolves. Wolves are the forebears of all present-day dogs. The point London was making (not very subtly at that) was that the wild brought out the "inner wolf" in a dog whose basic instincts had been crushed by humans; it was only because Buck's circumstances had drastically changed and the dog had become exposed to a more "authentic" environment were those instincts able to express themselves. The novel, which went on to sell two million copies, could easily be seen as sending a message to human readers who had also become too "domesticated."

Spencer and Sumner: To the Middle Class Belong the Spoils

Much of the credit for originally making social Darwinism so popular belongs to the British philosopher Herbert Spencer, who coined the phrase "survival of the fittest." Spencer (1820–1903) was already thinking about evolution before he read *Origin,* but Darwin's book convinced him that he was on the right track. In Spencer's version of Darwinian theory, though, the emphasis was placed on increasing freedom for individuals while limiting the role of government in dealing with social or economic inequities. Spencer was by no means alone in espousing this philosophy: The distinguished editor and economist Walter Bagehot championed social Darwinism in England, while the sociologist and economist William Graham Sumner took up the cause on the other side of the Atlantic. Both men believed that the process of natural selection would inevitably lead to the continuous improvement in the population.

Throughout the late nineteenth and early twentieth centuries, Sumner, who taught at Yale, wrote a series of essays expounding his belief in *laissez-faire capitalism* and individual liberty. The reason there were so many inequalities among men, he wrote, was because they were innate—people were born with different abilities and intelligence. Poverty resulted from inherent inferiority. Elaborating on his thesis, he declared that competition for property and social status—the "struggle for life," in other

Who Are They?

William Graham Sumner (1840–1910) was an American sociologist and economist who promoted the doctrine of social Darwinism. His most important book is *Folkways,* published in 1907.

What Does It Mean?

***Laissez-faire* capitalism** (from the French, meaning "leave alone") refers to an economic doctrine that says that the best economic system is one with no interference by government.

239

words—would lead to the elimination of the ill-adapted while simultaneously preserving racial soundness, morality, and cultural vigor. His view of morality was based on a middle-class Protestant ethic of hard work, thrift, and sobriety.

He wrote in his most famous book, *Folkways* (1907), that customs and morals arise in response to such basic instincts as hunger, sex, and fear. For that reason, folk customs were suspect and "inferior"; by stubbornly clinging to them, poor people made themselves resistant to reform. He deplored any effort on the part of the government to better the lives of the poor because the cost of such charity would fall on the middle class—his "forgotten man," as he called him. About 70 years later, in his presidential election campaign, Richard Nixon appealed to much the same constituency, which he called "the silent majority."

Dragging the Chain of Being out of the Closet

Social Darwinism also found fertile ground to take root in Germany. It's possible to make a case that the road that eventually led to Hitler many decades later began with what is known as the "biogenetic law." This "law" was the brainchild of the evolutionary biologist Ernst Haeckel. Haeckel, who has made his appearance in this story before, had conceived of the principle that "ontology recapitulates phylogeny"—the idea that the embryonic development duplicates the evolutionary development of the species. (This idea has little credibility among scientists, although some evidence can be found to support it in certain cases.)

The Sinister Implications of the Biogenetic Law

How could this questionable principle pave the way for Hitler? Let's see what Haeckel said. The biogenetic law proposed that development of more advanced species had to pass through stages embodied by adult organisms of more "primitive" species. Every step up the evolutionary ladder, then, was intended to set the stage for the appearance of man. It was an add-on process, in which one form of life evolved into the succeeding one by adding on a new feature. Humans, for instance, evolved when the embryo of the next highest life form—the ape—added on a new attribute in its embryonic development.

In contrast to Darwin's idea of evolution, which proposed a branching out of life forms (like a bush), Haeckel's was a one-way trip up a tree, describing a linear progression with a definite objective. Recall that, in Darwin's theory, natural selection does not act to bring about a particular "objective," nor is Darwin's evolution progressive by following a single direction. Humans are not "higher" than primates, as Haeckel believed, but rather are related by common ancestry.

But Haeckel didn't confine his theories to the realm of science. Instead, he tried to expand his ideas to society, with regrettable results. Haeckel's works reflect the influence of early anthropologists, who held that some cultures were "primitive" in the embryological sense. In other words, their development had stopped short of that of more advanced civilizations, such as Germany's. (We still talk about countries being "underdeveloped.") These anthropologists saw evolution in purely linear or hierarchical terms: Different races could be ordered from top to bottom of the evolutionary tree. There was no appreciation for the possibility that evolution was a far more complicated process or that all humans, regardless of race, are descended from the same group of hominids in Africa.

In 1902, Haeckel proposed a similar progression for rationality, with the lowest organic forms at the bottom and the greatest thinkers dominating the summit. By the same token, he believed that races differed in rationality as well, with the "savage races" at one end of the scale and Europeans at the other. The difference between the "savage" and "civilized" peoples, Haeckel stated, was greater than that between the savages and dogs. To Haeckel, that meant that the Europeans were right to subjugate the savage races. "The lower races … are physiologically nearer to the mammals—apes and dogs—than to the civilized European. We must, therefore, assign a totally different value to their lives." He didn't leave much doubt as to the ominous implications of his theory. Evolution was *kampf*—an eternal struggle in which the fit eliminated the unfit. (Hitler titled his autobiography *Mein Kampf*—"My Struggle.") Haeckel placed Jews, in particular, into the ranks of the inferior classes, arguing that Jesus wasn't a Jew at all, but rather the son of a Roman soldier who came from—where else?—Germany.

If Haeckel's insistence on pigeonholing peoples and races sounds familiar, it's Aristotle's ladder of life and the medieval great chain of being in another guise. The great chain of being was designed to celebrate the ascendancy of human beings, or, should we say, man? Because man is the gender that counts in Haeckel's scheme. Just as races were ranked higher and lower, so were the sexes. A further categorization—the only one that you had a chance of escaping—was age. That meant that white females were essentially put on the same rung of the evolutionary ladder as black men and European infants.

Carl Vogt

Such specious reasoning caused Carl Vogt, a professor of natural history at the University of Geneva, and a contemporary of Haeckel, to claim: "By its rounded apex and less developed posterior lobe, the Negro brain resembles that of our children, and by the protruberance of the parietal lobe, that of our females." Based on this purported anatomical correspondence, Vogt concluded that "the Negro brain" belongs "by the side of that of a white child." So does the brain of a white woman. Vogt goes on to say, "[T]he female European skull resembles much more the Negro skull than that of the European male." To back up his propositions, Vogt cited earlier studies, as questionable as his own, to construct his own version of evolution. "[I]n the Negro brain, both the cerebellum and the cerebrum, as well as the spinal cord, present the female and infantile European as well as the simious type." By *simious type*, Vogt means "simian" or "monkey-like." That is, blacks, women, and children were the link connecting apes to adult white males.

According to Haeckel, the same linear evolutionary system could be applied to the history of Western religion as well. Judaism recapitulated paganism, Haeckel asserted, and then transcended it. Christianity recapitulated paganism and Judaism, and then transcended both, in turn. In this conception, Judaism was embryonic Christianity, a more primitive form of religion, which set the stage for the development of the mature form.

What Does It Mean?

A **protist** is a group of simple organisms with characteristics of both plants and animals. Most protists are unicellar—composed of a single cell—and can be viewed only with a microscope.

A Dangerous Idea Lives On

Scientifically, Haeckel's biogenetic law couldn't hold up, but, while it was discredited, it retained a hold on the popular imagination. Even Dr. Benjamin Spock, to whom millions of Americans turned for instruction on how to raise their children, used Haeckel's ideas when he discussed the development of the human fetus. In the 1980s, a major American bank ran an ad in *Newsweek* depicting a path leading from a single-cell *protist* to a white, male, briefcase-toting banker. The accompanying text boasts that the bank has evolved into "a no-holds-barred, full-blooded, undistracted, single-minded bank for business." You can't get a much better expression of social Darwinism than that. Although this is obviously a caricature, it shows how insidious and powerful a myth can become.

The Virtue of War and the Subjugation of Savages

In its heyday, social Darwinism could be used to justify or explain practically any kind of war or colonization. After all, if one group or race was "superior" to another, the "superior" group was given license to do whatever it pleased. Of course, who was "superior" and who was "inferior" depended on your point of view. Both sides in the Franco-Prussian War of 1870, for instance, claimed that natural selection was operating in their favor. (If that were the case, then natural selection appeared to settle on the Germans, who drove the French forces all the way back to Paris in a matter of months.) This was also an era when the major European powers—Germany, France, Belgium, Portugal, and England—were engaged in the profitable business of colonizing much of Africa and Asia. Social Darwinism was particularly well suited as a rationale for the imperialist adventure.

In 1897, an influential American clergyman named Josiah Strong made the case for social Darwinism in unequivocal terms:

> The two great ideas of mankind are Christianity and civil liberty. The Anglo-Saxon civilization is the great representative of these two great ideas. Add to this the fact of his rapidly increasing strength in modern times, and we have a demonstration of his destiny.

He then goes on to predict that, while the Anglo-Saxons were among the most civilized "races" on Earth, they would have to fight to maintain their superiority in what he called "the final competition of races." But he had no doubt about the ultimate victor of such a struggle: "The United States will assert itself, having developed aggressive traits necessary to impress its institutions upon mankind. Can anyone doubt that the result of this competition will be the survival of the fittest?"

America wasn't immune from the temptations of social Darwinism, either. In many ways, President Theodore Roosevelt (1858–1919) was a confirmed believer in the theory. He expressed the opinion that a racial war to the finish with the Indians was bound to come sooner or later, and he advocated the idea that English-speaking people (presumably among the most civilized) should spread all over the world. He was convinced that only the powerful nations would survive and that the weak ones would die. To bring that result about, he advocated the virtues of war.

Roosevelt's sentiments were echoed by the German militarist Friederich von Bernhardi. In 1911, Bernhardi published an influential book, *Germany and the Next War,* extolling the virtues of war. War, he declared, was a "biological necessity," and its outcome "gives a biologically just decision because its decisions rest on the very nature of things." Any effort to peacefully mediate a conflict was a "presumptuous encroachment on the natural laws of development." According to Bernhardi, confirmation for this notion could be found by studying plant and animal life, which would prove that "war is a universal law of nature." His words didn't fall on deaf ears. Three years after the book's publication, Germany had plunged into World War I.

Who Are They?

Friedrich von Bernhardi (1849–1930) was a German officer, diplomat, and author who is best known as an apostle of German nationalism and expansionism. He believed that military might would give Germany the ability to exert dominance over Europe and control the world.

Let Free Markets Prevail

Social Darwinism also proved useful to support *laissez-faire* capitalism. This doctrine holds that an economic system functions best when there is no interference by government. It is based on the belief that unfettered economic activity is the most desirable state of affairs and that attempts by government to either regulate or stimulate free markets will disrupt the flow of goods and services. There is practically no system of *laissez-faire* capitalism in existence today, at least on a large scale. Even the United States, for all its emphasis on free markets, regulates and stimulates markets to some degree. For instance, the Securities and Exchange Commission sets certain standards for trading on the stock market.

In the mid-Victorian era—at a time when Charles Dickens was writing about the dismal conditions brought about by the Industrial Revolution in England—*laissez-faire* capitalism enjoyed supremacy. And social Darwinism came in handy as a theoretical framework to support its perpetuation. Advocates of the existing class structure, dominated by an elite with money and property, pointed out that the men on top (and they were mostly white men) deserved to occupy their position. That they were able to gain control of property and accumulate capital only testified to their superior and inherently moral nature. They had succeeded, social Darwinists said, because they were industrious and frugal, and because they practiced temperance—they didn't succumb to drink or drugs, vices associated with the poorer "inferior" classes.

That was why an economist such as William Sumner could deplore attempts to reform society through state intervention—it would be "unnatural" insofar as it interfered with economic freedom. (Of course, this freedom could be exercised only by the men with money and property to begin with.) The status quo, then, was merely a reflection of the "natural order"—an inevitable outcome of the struggle for existence in which the inferior classes were being gradually culled out to make room for the superior.

Mr. Marx, Meet Mr. Darwin

Interestingly, it wasn't only the defenders of capitalism who found in Darwinian theory a useful tool in promoting their own agendas. The founding fathers of communism, Karl Marx and Friedrich Engels, were also intrigued by Darwin's ideas and appropriated some of them in setting out their own theories. In December 1860, a year after Darwin had written *The Origin of Species,* Marx wrote to Engels, "Darwin's book is very important and serves me as a basis of struggle in history ...." They embraced the notion of "the struggle for existence" because it squared neatly with their own views. They were also seduced, as Haeckel was, by the concept of progressive evolution. In this case, though, it was social Darwinism turned on its head.

In the Marxist conception of history, history evolved through a blind process of class struggle and revolution. This struggle underwent a predictable sequence of stages: beginning with slavery in ancient times, followed by feudalism, which developed, in turn, into capitalism. According to Marx and Engels, the culmination of history would come about only with the triumph of communism. At that point, the laboring masses would rise up and overthrow the ruling class, creating as a result a "classless" society. Class, not race, religion, or gender, was the distinguishing categorization of this system. Karl Marx apparently felt so deeply indebted to Darwin that he was prepared to dedicate his book *Das Kapital* to him. Darwin declined the honor.

Who Are They?

Friedrich Engels (1895–1920) was a German revolutionary political economist who collaborated with Karl Marx on several influential analyses and critiques of capitalism. Together they are considered the founders of scientific socialism, better known to the world as communism.

Vestiges of Social Darwinism

As a phenomenon, social Darwinism has been stripped of much of its scientific and sociological credibility and has been exposed as a dangerous sham. Still, even now, it can't be said that the final nail has been hammered into its coffin. With vampire-like stealth, it continues to resurface, often in disguise, decked out in quasiscientific or sociological trappings. In 1994, for example, Richard J. Herrnstein and Charles Murray published a controversial book, *The Bell Curve,* in which they made the case that intelligence, race, and genes were all connected. Rightly or wrongly, many critics contended that the book's ideas, supposedly bolstered by rigorous sociological methods, were an attempt to justify the lower socioeconomic status of minorities. By implication, if racial or genetic traits were responsible for lower intelligence, then concerted efforts on the part of the government to redress social inequities were doomed. This is what you might call putting old wine in new bottles.

A Mississippi professor named Robert S. McElvaine summed up the whole matter succinctly when he told a recent convention of fellow historians, "Connecting biology with history is a practice that does not have a distinguished pedigree."

The Least You Need to Know

➤ Social Darwinism is a doctrine based on a distortion of Darwin's evolutionary theory to justify the existence of social, economic, racial, religious, and sexual inequalities.

➤ Social Darwinism sees evolution as a progressive system (which it is not) in which peoples, groups, races, religions, and so on all have a fixed place—some supposedly "superior" and others "inferior."

➤ In England, the leading proponent of social Darwinism was the philosopher and economist Herbert Spencer; in the United States, his counterpart was William Sumner, also an economist, who held that the highest values— industriousness, frugality, and morality—were represented by the middle class.

➤ In Germany, the foremost advocate of social Darwinism was Ernst Haeckel, an evolutionary biologist who conceived of an evolutionary theory stating that white male supremacy was what was intended by nature.

➤ Social Darwinism was used to justify the status quo—those who had money and property were the most "fit" and thus deserved their privileges.

➤ Although social Darwinism has been largely discredited, it persists in one form or another, as evidenced by the recent publication of books that attempt to explain why racial inequities in the United States may be connected to genes and innately lower intelligence.

Chapter 22

Eugenics: The Dangers of Purity

In This Chapter

➤ Francis Galton and the debut of eugenics

➤ Crusading for eugenics in Britain and America

➤ Germany's embrace of eugenics

➤ Sociobiology—is social behavior genetically determined?

➤ E. O. Wilson: why racial differences matter

As we saw in the last chapter, social Darwinism was eagerly adopted by the rich and the propertied (and the white European males)—and why not? The theory offered them the reassurance that their success was conferred on them by natural election. They were the "fit," and people who were poor, mentally or physically disabled, or prone to unspeakable vices were clearly "unfit." "Fit" also implied intellectual, moral, and physical excellence. (It rarely seems to have occurred to social Darwinists that someone could be poor and a genius.) But supporters of social Darwinism perceived dangers on the horizon: It wouldn't be long before the poor, who had the irritating habit of reproducing in ever-greater numbers, were threatening to upset the balance of the population—a "dumbing down" on a national scale.

Science offered a "solution" to the problem: eugenics. The word, first coined by Sir Francis Galton (who was Charles Darwin's cousin), referred to a doctrine calling for the "improvement" of the population by means of influencing heredity. This could be done, Galton said, by encouraging the breeding of those with the superior traits (exemplified by success, presumed moral uprightness, and higher intelligence) while curtailing the breeding of the poor and less intelligent. Although the doctrine was based on bad science, Galton and his followers marshaled a large body of data intended to corroborate their theory. Eugenics was soon embraced in both Britain and the United States, and was used to justify the involuntary castration and sterilization of people judged to be unfit for whatever reason.

What Does It Mean?

Eugenics is a doctrine born in the nineteenth century that claims that a society's heredity can be controlled by encouraging the "purebred" upper classes to have more children while curtailing the ability of the poor, the mentally ill, or criminals to reproduce.

The greatest eugenics experiment of all time, however, took place in Nazi Germany, beginning in the late 1930s, when millions of people were slaughtered because they were declared racially, mentally, or physically inferior. As a result of their atrocities, eugenics now carries the stench of the concentration camp. Some scientists, known as sociobiologists, maintain, however, that the sordid legacy of eugenics shouldn't be allowed to stop them from investigating the differences in heredity among various peoples and races. All humans constitute one species, but sociobiologists say that there are some small genetic differences just the same and that it would be wrong to ignore them. Predictably, their efforts have sparked heated protests from critics who contend that any attempt to link evolution and behavior is misguided and can also have dangerous consequences, as history shows all too well.

Breeding Out the "Undesirables"

Eugenics first entered the English vocabulary in 1883. The man responsible for its inauspicious debut was Sir Francis Galton (1822–1911), an Englishman who also happened to be a cousin of Charles Darwin. The word comes from the Greek *eugenēs,* meaning "well-born." Eugenics was basically a method of putting social Darwinism to work in practical terms. Galton and the eugenicists who came after him believed that human heredity could be regulated to enlarge the population of the prosperous and superior classes while curtailing the ability of the lower classes to breed. The idea, simply stated, was to enhance the "desirable" traits in a population through reproduction and to diminish the number of "undesirable" traits.

Galton's use of a Greek word to coin eugenics is no surprise. Galton loved the ancient Greeks, especially the Athenians, whom he believed had a civilization second to none. In his book *Hereditary Genius* (1869), he attributed the Athenians' accomplishments in art, literature, and philosophy to a "system of partly unconscious selection." The Athenians benefited from a caste system, maintained by slavery, that perpetuated the ranks of the well-born, uncontaminated by traits from outsiders. Eventually, though, Athens declined. Galton thought he knew why: Too many foreigners "of a heterogeneous class" were coming into the city. Among the "pure" Athenians, morality was deteriorating and marriage was no longer held in high regard. Soon the purebred Athenian became an endangered species. To be sure, Galton's reading of ancient history is too idiosyncratic for most historians to accept, but to Galton, it offered a cautionary lesson: Humans had the power to improve or to cause great damage to their own species.

Panic of the Ruling Class

Galton feared that his England, just like ancient Athens, was in danger of decline as a result of a population growth of the lower social classes. The poor would continue to bring ever-greater numbers into the world, while the well-off purebreds left relatively few offspring. In *Inquiries into Human Faculty,* published in 1883, he complained that the very people who should be having children were the same ones with the foresight and self-control to delay marriage.

Who Are They?

Sir Francis Galton (1822–1911) was a British scientist and cousin of Charles Darwin who introduced the idea of eugenics. Among his books are *Hereditary Genius* (1869), *Inquiries into Human Faculty* (1883), and *Natural Inheritance* (1889).

This was exactly the opposite of what should be happening, Galton argued. Social distinctions, he was convinced, reflected differences in innate endowment—the social Darwinist line. Like the ancient Athenians, the middle and upper classes possessed more "civic worth" than the lower classes. If you were successful, it was because you had inherited the ability to be a success from your parents. The only way of ensuring that the intellectual standards of the nation would rise, Galton thought, was by encouraging the early marriage and reproduction of these "thriving families." Eugenics, then, was nothing less than a program for national survival.

Statistics in the Service of "Purity"

Like his celebrated cousin, Galton was blessed with a wide-ranging curiosity, but he reserved a special passion for mathematics, particularly for a new branch of mathematics called statistics. "Whenever you can, count" was Galton's motto. The age of quantification was at hand in both Britain and the United States. The Industrial Revolution had brought with it the increased need to measure and survey a tidal wave of data. Statistics seemed to offer a reliable way of arriving at the inherent truth underlying natural phenomena. As far as Galton was concerned, the timing couldn't be better. By incorporating his theory of genetic purity within a statistical framework, he could give eugenics a scientific respectability that it might not otherwise enjoy. To that end, he became absorbed in the study of the mathematical distribution of what he called "natural ability" among a sample of British subjects.

Galton pored through biographical sketches of eminent Englishmen looking for evidence of natural ability. To his delight, he discovered that many of these distinguished figures were related to one another. That only provided further evidence that intelligence and talent must be due to heredity. Seeking additional corroboration for his theory, Galton collected statistics on the height, dimensions, and strength of large numbers of people, calculating correlations between various pairs of attributes. "Could not the undesirables be got rid of and the desirables multiplied?" he wondered.

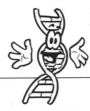

A Natural Selection

Statistics, in a primitive form, have been used for thousands of years. Before 3000 B.C.E., Babylonians recorded tabulations of agricultural yields and the sale of goods on clay tablets. But it was only in the nineteenth century that statistics were recognized as a branch of mathematics to quantify information that could not be verbally conveyed with precision.

Yet, Galton was on to something that had escaped his cousin. For instance, Galton used the word *genius* to refer to "an ability that denoted exceptionally high intelligence but was also "inborn." Both mental and physical attributes, he wrote in his book, *Hereditary Genius,* could be inherited. It was an unusual idea for the time. Even Darwin hadn't quite realized that intelligence was an inheritable attribute. Once he read Galton's book, he wrote back to him: "You have made a convert of an opponent in one sense, for I have always maintained that, excepting fools, men did not differ much in intellect, only in zeal and hard work." When Darwin wrote *The Descent of Man,* published in 1871, he was at least partly inspired by Galton's views.

To be fair, Galton was not in favor of the creation of an aristocratic elite; he envisioned an entire population made up of superior men and women. And he balked at radical solutions, preferring to encourage more "good" marriages rather than restricting "bad" marriages. All the same, the damage was already done—a dangerous idea had been put into circulation.

Introducing Eugenics to the World

On May 16, 1904, Galton was ready to present his ideas to the public. At a meeting of the Sociological Society (founded the previous year), he put forward his views in a paper titled "Eugenics: Its Definition, Scope, and Aims." In it Galton urged the society's members to launch a crusade to make eugenics a national religion. One of the first to express his enthusiasm for the plan was the playwright George Bernard Shaw. Shaw believed that a society based on eugenics was far better than one in which men and women "select their wives and husbands less carefully than they select their cashiers and cooks." But not every member of the Sociological Society was as prepared to embrace eugenics as Shaw was. The cool reception that greeted his paper convinced Galton to form his own organization—the Eugenics Education Society—in 1907.

"Three Generations of Imbeciles Are Enough"

Interest in eugenics only grew when Mendel's laws of inheritance were rediscovered in the early 1900s. Now that the role of dominant and recessive traits was recognized, eugenicists felt that they would be in a better position to prove their belief that natural ability could be inherited. In the United States, biologist Charles Davenport (established a center for research in human evolution at Cold Spring Harbor, New York. A confirmed Mendelian, Davenport believed that single-unit genes could be found for such undesirable traits as alcoholism and feeblemindedness. Prostitution, he maintained, was not caused by poverty but by "innate eroticism." So, if you could prevent reproduction by people who had these traits, he believed, you would eventually eliminate these traits from the human population.

Sterilizing the "Undesirables"

It was his hope that "human matings could be placed upon the same high plane as that of horse breeding." This goal, he asserted, would require drastic measures—eugenic castration, for one. If castration seemed too extreme, sterilization did not. Between 1907 and 1937, some 32 states required sterilization of people judged as "undesirable," a category that included the mentally ill, the handicapped, those convicted of sexual or drug-related crimes, and others who were lumped together as "degenerate."

Who Are They?

Charles Davenport (1866–1944) was an American zoologist and a leading proponent of eugenics who advanced the use of statistical measurements in biological research.

Throughout the first decades of the twentieth century, eugenicists advanced their doctrine with messianic fever. In 1906, the Committee on Eugenics was founded in the United States for the purpose of educating Americans about the value of superior blood and the menace to society of inferior blood. In both the United States and Britain, the public flocked to fairs and exhibitions featuring exhibits filled with photos of guinea pigs and portraits of men and women that were supposed to demonstrate characteristics unique to the criminally insane or the mentally unbalanced.

Even the distinguished American Museum of Natural History in New York got on the bandwagon, hosting the 2nd International Inhabitation of Eugenics in 1921. The text that accompanied one of the exhibits on display exhorted visitors: "Unfit human traits such as feeblemindedness, epilepsy, criminality, insanity, alcoholism, pauperism, and many others run in families and are inherited in exactly the same way as color in guinea pigs." Young couples were advised to pay close attention: "If All Marriages were Eugenic, We Could Breed Out Most of This Unfitness in Three Generations." Nor should people fall victim to complacency or suffer pangs of conscience when it came to rooting out inferior traits from the population. "Are we to be so careful of our pigs and donkeys and cattle—and then leave the ancestors of our children to chance or 'blind' sentiment?"

Seduced by Eugenics

Not everyone bought into the theory of eugenics. Several notable public figures, including British essayist G. K. Chesterton, American journalists H. L. Mencken and Walter Lippmann, and lawyer Clarence Darrow, made no secret of their contempt for the whole business. Unfortunately, many intellectuals, businessmen, and political figures felt differently—among them were Winston Churchill, Alexander Graham Bell, Theodore Roosevelt, John Maynard Keynes, and the usually taciturn Calvin Coolidge. While Coolidge was vice president, he declared, "Nordics deteriorate when mixed with other races." His statement only reflected the anti-immigrant sentiment that was taking hold in the United States at the time. America put out the welcome mat for Nordic immigrants, but others were shunned. In 1924, the Immigration Restriction Act was enacted to curtail admission to the country of people from countries referred to as "biologically inferior."

Who Are They?

Oliver Wendell Holmes Jr. (1841–1935) was one of the leading legal authorities of his time and served as an associate justice of the U.S. Supreme Court for three decades from 1902 to 1932.

Nine Justices vs. One "Moral Imbecile"

In 1927, the U.S. Supreme Court had the chance to rule on the validity of eugenics laws passed by the states. The case Buck v. Bell, arose as a result of an appeal of Virginia's decision to sterilize Carrie Buck, an institutionalized 17-year-old whom the state had decreed a "moral imbecile." Carrie was the daughter of a "feebleminded" mother and the mother herself of a daughter who, at age 7 months, was found to have below average intelligence. The court ruled 8-to-1 to reject Buck's appeal. In his majority opinion, Oliver Wendell Holmes wrote, "Three generations of imbeciles are enough."

Why was the eugenics movement so popular? One reason—perhaps the most powerful reason—was fear. Historian Daniel J. Kevles, author of *In the Name of Eugenics* (1985), suggests that "eugenicists identified human worth with the qualities they presumed themselves to possess—the sort that facilitated passage through schools, universities, and professional training." In the United States, most of these academicians were white Protestants of European descent—Wasps—who feared that their lives and those of their children were jeopardized by waves of darker-skinned Catholic and Jewish immigrants from southern and eastern Europe.

The Ultimate Eugenics Experiment

Although eugenics caught on in the United States and Britain, it never had so much influence as it did in Germany. The twentieth century wasn't very old before Dr. Alfred Ploetz founded the German Society of Racial Hygiene with the objective of improving the hereditary qualities of the population. In the 1920s, German textbooks were instructing students in the finer points of heredity and racial purity. By 1933, shortly after Hitler came to power, eugenics became the law of the land. Formally known as the Eugenic Sterilization Law, it mandated sterilization "for the prevention of progeny with hereditary defects" in cases of "congenital mental defects, schizophrenia, manic-depressive psychosis, hereditary epilepsy … and severe alcoholism." Even blindness and physical deformities, judged as "objectionable," could make an afflicted individual liable for sterilization.

When the victims protested, they were assured that their personal sacrifices would serve the common good. "We go beyond neighborly love," said one official. "We extend it to future generations. Therein lies the high ethical value and justification of the law." Some 350,000 schizophrenics and other mentally-ill Germans were involuntarily sterilized, along with 30,000 German gypsies as well as a few hundred black children.

The attempt to put eugenics into practice didn't stop with sterilization: Over the next several years, the Nazis went on to kill 70,000 psychiatric patients. In 1939 alone, 300,000 disabled people were gassed at Auschwitz. While the initial eugenics laws were not directed against Jews, the sterilization of "undesirables" laid the groundwork for the genocide to come. What began as an immoral racial hygiene campaign would culminate in the Holocaust.

The "Final Solution"—the deliberate extermination of the Jews (as well as 750,000 gypsies and several thousand homosexuals)—was the horrifying realization of Francis Galton's vision of human improvement. Understandably, the eugenics movement never recovered from the atrocities committed in its name by the Nazis. But eugenics wasn't dead, by any means. On the contrary, it is still alive and well in some parts of the world. In 1995, for example, the Chinese promulgated the Law on Maternal and

Infant Health Care, requiring premarital screenings to determine whether either partner carries "genetic diseases of serious nature"; infectious diseases such as AIDS, gonorrhea, syphilis, and leprosy; or a "relevant mental disease." If either the man or woman has one of these conditions, the law stipulates that marriages will be permitted only after the couple has been sterilized. In its defense, a health minister cited statistics showing that China "now has more than ten million disabled persons who could have been prevented through better controls."

Who Are They?

Frederick Osborn (1889–1981) was a leading advocate of a reborn eugenics movement in the United States. He served as a brigadier general in World War II and was a fellow of the American Association for the Advancement of Science.

What Does It Mean?

Sociobiology is a developing field of science that investigates behavioral patterns among different groups, races, and populations of animals (including humans) in the belief that many types of behavior can be inherited.

What's in a Name?

In a more benign form, eugenics still exerts an influence on the United States. After World War II, a leading eugenicist, Brig. Gen. Frederick Osborn, advocated the establishment of "heredity clinics," which, he said, were "the first eugenic proposals that have been adopted in a practical form and accepted by the public." Osborn recognized, however, that he had a public relations problem on his hands. "Eugenic goals are most likely to be attained under a name other than eugenics." The word, he realized, had to go. Taking Osborn's words to heart, the *Eugenics Quarterly* changed its name to *Social Biology*. The same year, the English journal switched from *Eugenics Review* to the *Journal of Biosocial Science*. Then the societies changed their own names: The American Eugenics Society renamed itself the Society for the Study of Social Biology, and, in England, the Eugenics Society became the Galton Institute.

Sociobiology: A Successor to Eugenics?

The field of social biology, known today as *sociobiology,* is still a work in progress, and its acceptance is by no means universal. Essentially, sociobiology is a field that seeks to apply natural selection to the social behavior of animals, including humans. Such social phenomena as aggression, altruism, territoriality, and the ways in which mates are chosen are all considered grist for its mill. Just as Darwinism proposed that natural selection acts by selecting for or against different traits, sociobiologists believe that it selects for or against different behaviors. According to the theory, some behavioral patterns will emerge and then spread in a population, while others will disappear completely.

The Ghost of Eugenics?

Sociobiology has come under fire because some critics oppose any attempt to link evolution with human behavior. As evolutionary anthropologist Dr. Robert Boyd put it: "There are a whole bunch of people who think it's dehumanizing to talk about humans in biological terms." Critics fear that sociobiology promotes biological determinism. If patterns of behavior are hereditary, they say, then that would imply that behavior is fixed in the genes, implying that some groups of people, for instance, might be naturally "passive" or "athletic" or "intellectual." For these critics, the ghost of eugenics is still stalking the ramparts. In addition, sociobiology is seen as supporting the existing social structure—charges that echo those leveled against social Darwinians more than a century ago.

Sociobiologists counter that heredity doesn't dictate all behavior—far from it. Inheritance is a grab bag of talents, temperaments, intellectual capacity, and so on. How we use that inheritance is up to us, sociobiologists say. We learn to acquire language instinctively, but how well we express ourselves is within our control. By the same token, we are attracted to certain people, but whether we act on that attraction is a voluntary decision.

Is Selfishness the Best Strategy for Survival?

Where sociobiologists have had the most success, at least in scientific terms, is in the investigation of the evolution of social behavior patterns in insects, such as ants and bees. These insects thrive in very well-organized societies, where the survival of individual members of the group is of less importance than the survival of the group as a whole. As we noted earlier in our story, natural selection does not favor altruism unless an act of generosity on the part of the donor is later rewarded by a reciprocal act on the part of the recipient. Otherwise, say the Darwinians, by acting altruistically, without any certainty of a payoff later, you'd be sacrificing your reproductive edge.

Yet biologists have observed many instances in which members of a group will do favors for one another, even though the donor sacrifices its ability to have offspring. Even Darwin recognized the problem. How selection can reward such sacrifice was unclear to him. As time went on, he thought, the traits for altruism should become uncommon and then disappear altogether because their possessors would be reproducing less often than "selfish" individuals.

In 1960, British biologist W. D. Hamilton developed the concept of kin selection, which showed that individuals within a species can best enhance their own reproductive success by aiding their close relatives. The only qualification was that the gain conferred on the recipient had to be much greater than the cost to the donor. The concept of reciprocal altruism even appears to apply in situations in which no close genetic relationship exists. Another biologist even showed how species that started out selfishly could develop into altruistic beings over time and enjoy higher reproductive success as a result.

Sociobiology's Leading Advocate: E.O. Wilson

Several theories have been proposed to account for all the ways in which individuals and groups belonging to various species behave in response to the pressures of natural selection and how they deal with their environment. All these theories were finally brought together by the world's foremost sociobiologist, E.O. Wilson, in his 1975 book *Sociobiology: The New Synthesis*, regarded as one of the seminal texts of the field. If Wilson had limited his subject to an examination of the behavior of ants, for which he has won great renown, his book would never have set off the controversy it did. Well aware of the risks he was taking, Wilson plunged ahead and tackled the sensitive issue of the differences in race, making sure, though, to note that he wasn't trying to suggest some kind of eugenic approach to the subject. This is how he explains his view:

> Given that humankind is a biological species, it should come as no shock to find that populations are, to some extent, genetically diverse in the physical and mental properties underlying social behavior. A discovery of this nature does not vitiate the ideals of Western civilization. We are not compelled to believe in biological uniformity in order to affirm human freedom and dignity.

Genetic diversity, he declared, was well worth studying. We shouldn't look on the legacy of genetic diversity as a source of despair, he went on, but rather regard it with "hope and pride ... because we are a single species, not two or more, one great breeding system through which genes flow and mix in each generation." He went on to say:

> Because of that flux, mankind viewed over many generations shares a single human nature within which relatively minor hereditary influences recycle through ever changing patterns, between the sexes and across families and entire populations.

Who Are They?

Edmond Osborne Wilson (1929–) is an American biologist and entomologist who is the world's leading authority on ants. He is a leading advocate of sociobiology, a field of science that studies the social behavior of all animals, including humans.

But it made no sense, Wilson argued, to pretend that there aren't *any* differences among humans and that some of the different ways that people act are based on physical and mental traits that they inherited. Sociobiology obviously still has a hard road to hoe because the territory that it means to survey has already been so contaminated by the legacy of eugenics and social Darwinism. As we've seen, the study of biological diversity is an undertaking that can be carried out for the best—and worst—of motives.

The Least You Need to Know

➤ In 1885, Sir Francis Galton, an English scientist (and Darwin's cousin), introduced a doctrine called eugenics, intended to "improve" the English population.

➤ Eugenicists maintained that "superior" individuals should be encouraged to marry and reproduce and that those who were considered "inferior"—the poor, the mentally or physically disabled, criminals, or people prone to violence, alcohol, or drugs—should be discouraged from reproducing.

➤ The eugenics movement was enthusiastically adopted by many intellectuals, scientists, and businessmen on both sides of the Atlantic.

➤ Eugenics captivated the voters in America, too; several states enacted laws imposing sterilization on people judged retarded, mentally ill, or criminally insane.

➤ The Nazis carried the doctrine of eugenics to an extreme by sterilizing and later executing people condemned as unfit or inferior.

➤ In spite of the taint of eugenics, serious scientific attempts, notably in the field of sociobiology, are still being made to study inherited differences among populations to see how biological diversity might influence social behavioral patterns.

Chapter 23

Creationism: Putting God Back into the Picture

In This Chapter

➤ Introduction to creationism

➤ Fundamentalist attitudes toward evolution before World War I

➤ The crusade against teaching of evolution after World War I

➤ William Jennings Bryan and the Monkey Trial

➤ Creationism today

When the theory of evolution was first introduced in the middle of the nineteenth century, it caused a huge uproar in scientific circles. But, with very few exceptions, scientists now accept Darwin's theory as the most accurate description of the origin and development of life on Earth that has yet been devised. This isn't to say that scientists are in agreement about all aspects of the theory. Debates continue to flare up about a variety of issues that are far from resolved—punctuated equilibrium, sociobiology, and the origins of the human species, to name just three. But there is no serious dispute among scientists about the basics of Darwinian theory: the idea that species can change over time as a result of Natural Selection or that all existing species have a common line of descent.

The general public, however, has been far less welcoming to Darwinism. Many Americans believe that evolutionary theory is opposed to the biblical teaching that God had a significant role in creation. One reason that the theory of evolution has run into such a groundswell of opposition from some (but hardly all) religious organizations can be traced to some degree to the cultural divide between scientists and the rest of the population.

A 1999 national survey by *Scientific American* magazine, for instance, found that fewer than 10 percent of the members of the prestigious National Academy of Sciences believe in God, in contrast to the 90 percent of Americans who say they do. However, as you'll discover, evolutionary theory is not necessarily incompatible with religious beliefs.

Another problem, of course, is the question of how life originated. Even if our common ancestor turns out to be a single-cell organism, as evolutionary biologists maintain, how did that organism get here? Why was there nothing one day and something the next? The question goes to the heart of a debate that has been going on for over a century in the United States and that is being played out on many levels: on Capitol Hill, in statehouses, in churches, in school boards, and on college campuses.

Joining the Debate over Evolution

Who are the participants in the debate over evolution? It would be simplistic to say that there are only two sides. Not all scientists who support evolutionary theory, for instance, feel that the origin of life is a topic to which science has anything useful to contribute. And many people who are deeply religious have no trouble supporting Darwin's account of evolution. Even Darwin took pains to reassure readers that his theory did not mean to exclude God. However, for the sake of convenience, we can generally say that the participants in the debate include scientists on one side and creationists on the other. *Creationism* is not a single philosophy or belief, nor does it invariably refer to a literalist interpretation of Genesis, in which God is seen as creating the universe in seven 24-hour days. Basically, we can define creationism as a range of beliefs, all of which attribute the creation of life to divine intervention. Adherents of this position are known as creationists.

What Does It Mean?

Creationism is a belief that the origin of life on Earth and all existing species is attributable to some form of divine intervention.

Some theorists who sympathize with the creationist view have been edging away from a more traditional and literal view of divine intervention and have staked out a position based on the very complexity of

260

organic life. The "irreducible complexity" of life, these theorists say, couldn't have arisen by chance alone—the blind action of natural selection at work, which is what Darwinians believe. Instead, it shows evidence of a divine hand at work—an "intelligent designer." The theory of *intelligent design* actually has won over some scientists to its side, although most remain adamantly opposed to it.

But before we consider traditional creationism and intelligent design in detail, we first need to see how we got to this point. Ironically, in England, where evolution was first introduced, the passions ignited by Darwin's theory have long since ebbed. But in the United States the debate has, if anything, grown only more contentious and bitter over the years. To see why so much skepticism about evolution has persisted in the United States, we have to go back to the beginnings of the twentieth century.

What Does It Mean?

Intelligent design is a theory based on the belief that because life is so complex, it couldn't have come into existence as a result of chance (as Darwinian theory states). Instead, this complexity shows evidence of a cosmic or intelligent designer, which may or may not correspond to the God of the Bible.

America After Darwin and Before Creationism

In the early years of the twentieth century, Protestant evangelical movements were actually far less critical of Darwin's theory than they are today. No organized campaign, for instance, was mobilized to ban discussion of the theory in classes before World War I. To be sure, evolution was criticized for departing from the story of creation in the Bible, but it wasn't considered particularly threatening to Christian doctrine. In an essay written for *Fundamentals,* an influential evangelical publication, George Frederick Wright, a minister and geologist, even went so far as to agree that the Bible "taught an orderly progress from lower to higher forms of matter and life." But he hastened to add that the first humans could come into existence, as the Bible specified, "by the special creation of a single pair, from whom all the varieties of the race have sprung."

It wasn't until after World War I that fundamentalists began to look upon evolution as the enemy. Their change of heart was partly a result of the war itself. The social Darwinist notion of the survival of the fittest had taken on new meaning in light of German aggression. In fact, many fundamentalists believed that the influence of Darwin's theory of evolution had helped spark the war in the first place. The fundamentalist who took up the crusade against evolution was a Presbyterian layman and three-time presidential candidate named William Jennings Bryan.

The Monkey Trial

William Jennings Bryan was not a literalist, though—that is, he didn't subscribe to the idea that creation took place only in seven literal days or that the Earth was a little more than 5,000 years old, as the seventeenth-century Irish Archbishop James Ussher had maintained. Bryan followed what was called the day-age theory, which took the view that the "days" of creation were metaphorical, implying that creation was a long process that might have taken hundreds of millions of years. Bryan's belief in the day-age theory was held by most of the leading creationists of the 1920s. (Seventh-Day Adventists were the most significant exception, believing that the Earth's history could be encompassed in only 6,000 years.)

Who Are They?

William Jennings Bryan (1860–1925) was a leading political figure, editor, and lecturer who ran unsuccessfully three times as a Democratic candidate for president—in 1896, 1900, and 1908. A spellbinding orator, he spearheaded a fundamentalist campaign against evolution that led to his participation in the famous Scopes trial of 1925.

Exploiting the oratorical skills that had propelled him into the national spotlight as a presidential candidate, Bryan was able to mobilize considerable opposition to the teaching of human evolution in schools. In the 1920s, three states—Tennessee, Mississippi, and Arkansas—passed laws banning evolution from the classroom, and Florida and Oklahoma officially condemned it. Many schoolteachers and civil libertarians were outraged by these restrictions.

Testing the Law Against Teaching Evolution

In 1925, a Tennessee high-school physics teacher and football coach, John T. Scopes, volunteered to test the statute with the assistance of the American Civil Liberties Union (ACLU). Ironically, Scopes wasn't even sure that he had actually ever taught evolution! The law that he proceeded to break specified that it was forbidden to teach that humans had been created in any other way except as put forward by the Bible. The stage was set for what would become known to history as the "Monkey Trial."

The trial attracted reporters from all over the world—it was one of America's first media circuses. The Scopes trial was never a truly criminal trial, nor was Scopes in any jeopardy even if he was convicted. The textbook that Scopes had used to break the law was called *A Civic Biology,* by G. W. Hunter. Hunter's bigoted ideas of evolution would prove offensive today—and not just to critics of evolutionary theory. A confirmed eugenicist, Hunter said that humans had evolved from simple forms of life, with Caucasians being "firmly the highest type of all." Blacks were considered the lowest. He was writing at a time when no less than 35 states had passed eugenic laws requiring the sexual segregation and sterilization of people regarded as "unfit" because of mental illness, retardation, or epilepsy. If such people were "lower animals," Hunter wrote, "we would probably kill them off to prevent them from spreading." Seeing that these were humans, he admitted that asylums were the next best solution.

When the grand jury convened in Dayton, Tennessee, to decide whether Scopes should be indicted, presiding Judge John T. Raulston left no doubt where he stood on the issue. No sooner had he finished reading the statute that Scopes was charged with violating than he proceeded to read the entire first chapter of Genesis. The grand jury obligingly returned an indictment, although it later was revealed that the jury's foreman believed in evolution. However, the only question before the grand jury was whether Scopes had broken the law and, therefore, should be indicted. Whether the law was constitutional was not an issue.

Evolution Goes Up on Trial

The trial began on July 10. Defending Scopes was Clarence Darrow from Chicago, who was one of the greatest trial lawyers of the century. In spite of his reputation, the ACLU objected to Scopes hiring him on the grounds that Darrow would pursue a grandstanding defense rather than construct a carefully reasoned argument. The ACLU was right to worry. Darrow's strategy was to portray the state authorities and their defenders as bigoted ignoramuses rather than to win an acquittal. His real objective was to turn the trial into a conflict between the forces of modernism and forces of a regressive fundamentalism. In that sense, Darrow was preaching not to the jury, but to the world.

Huge crowds gathered outside the courthouse as journalists and radio reporters jockeyed for space inside. Finally, the judge simply moved the proceedings out onto the courthouse lawn so that everyone could get a chance to watch. As the trial continued in the sweltering summer heat, Darrow and Raulston clashed over which witnesses would be admitted. When the judge ruled against scientists that Darrow had summoned from Harvard and the University of Chicago, they stepped out on the lawn and delivered their views to the public instead. Taking the stand as a witness for the prosecution, Bryan proved no match for Darrow. The attorney methodically attacked Bryan's fundamentalist views—Darrow believed that Christianity was a "slave religion," to begin with—while exposing the former candidate's ignorance of biology. As one of his exhibits to prove evolution, Darrow presented "Nebraska Man"—remains which were then thought to be of an extinct prehuman but that turned out to belong to an extinct pig.

Who Are They?

Clarence Seward Darrow (1857–1938) was one of the best-known American defense lawyers. His fame rests principally on two cases: the Scopes trial and the Leopold-Loeb trial, in which he convinced a jury to spare two admitted murderers from execution.

Darrow knew that he was going to lose the case, but he was unconcerned about the verdict because he intended to seek a reversal on appeal to the state Supreme Court, the forum that he was seeking all along. With nothing to lose in Dayton, he mocked both judge and jury. So, he actually asked the court to instruct the jury to find Scopes guilty as charged. (He feared a hung jury, which would have risked a retrial.) Throughout the trial, Raulston ruled out any attempt on Darrow's part to turn the trial into a debate over the validity of Darwin's ideas or the constitutionality of the statute. The only issue was whether Scopes had violated the law, and this the jury found that he had done. The judge then imposed a $100 fine on Scopes.

Making Sense of the Monkey Trial

Much to his frustration, Darrow never had a chance to appeal—the state Supreme Court overruled the verdict on technical grounds, eliminated the fine, and ordered the prosecution dismissed. The contested law stood for another 42 years, and Tennessee's textbooks remained mute on the subject of evolution. (The restriction wasn't limited to Tennessee: Until 1974, publishers expunged references to Darwinian theory all across the nation, to avoid potential legal problems and to forestall threats to sales.) The trial marked Bryan's last hurrah. Broken and ill, he died suddenly on July 26, five days after the end of the trial that probably had hastened his end. Forgotten amid the controversy was the fact that Bryan had taken a courageous stand against the brutal eugenics advocated by the author of the disputed textbook.

The Monkey Trial had a second run in 1960—this time on the big screen. In the movie version, *Inherit the Wind,* Spencer Tracy plays the Darrow role and Frederick March enacts the role of the prosecuting attorney who takes the position that the biblical account of creation is in conflict with teaching of evolution. The movie wasn't really true to history, though. In retrospect, the trial was more about clashing egos than it was about the validity of evolutionary theory. As Harvard paleontologist Stephen Jay Gould wrote, "The usual reading of the trial as an epic struggle between benighted Yahooism and resplendent virtue simply cannot suffice—however strongly this impression has been fostered."

Creationism Comes of Age

In 1961, the Seventh-Day Adventist teacher and amateur geologist George McCready Price published a book called *The Genesis Flood,* in which he proposed that most of the fossil-containing rock formations could be explained by the biblical flood. The book, a collaborative effort with a conservative biblical scholar and a hydraulic engineer, represents a throwback to the early catastrophists who attempted to connect most major geological changes to the Deluge.

The book gained widespread popularity after Price's death in 1963 and led to the formation of the Creation Research Society. The society promoted a doctrine known as young-earth creationism, which, as the term suggests, is centered on the belief that the history of Earth required only a relatively brief span of time—10,000 years—to take place. (The majority of creationists accept day-age and other theories that allow for a much longer time frame, stretching into the hundreds of millions of years. Geological research conducted in 2001 in western Australia has found mineral evidence for the existence of continental crust and oceans 4.4 billion years ago.)

At the beginning of the 1970s, followers of flood geology set their sights on airing their views in public schools. To make their theory more acceptable to a secular society, they took out the biblical references and called their theory scientific creationism or creation science. And rather than trying to ban the teaching of evolution, creation scientists argued instead for equal time for their views. The attempt to put creationism on an equal footing with evolution met with some success. In the early 1980s, Arkansas and Louisiana passed laws mandating the teaching of creation science along with evolutionary theory. In 1987, though, the U.S. Supreme Court dealt the creationists a setback by declaring such laws a violation of the Constitutional restrictions governing the separation of church and state. Undaunted, creationists began to concentrate on winning representation on local school boards, where they would have more control over the curricula.

Creationists Have Their Say

Before we move on to the latest developments in the battle over Darwin, it might be helpful to take a more detailed look at how creationists present their views and why they object so passionately to evolutionary theory. What follows is derived from an essay titled "Science, Creation and Evolution," by Babu G. Ranganathan, published by the Institute for Creation Research, a branch of creationism known as *Young Earth Age* (YAC).

Confusing the What with the How

First, creationists say, science can deal only with *how* the universe works because that's all we can actually observe and test. That means that the subject of the origin of life and the universe is outside the scope of human observation and, therefore, cannot be investigated by science. Because no human was present to observe the universe coming into existence, both evolution and creation are ultimately positions of faith, not science. Where evolution can be put to the test is by evaluating the actual empirical facts of science. Creationists believe that if all the scientific data—gathered from genetics, physics, paleontology, mathematics, and so on—are properly examined they will "support faith in ... the belief that an intelligent power

What Does It Mean?

Young Earth Age refers to the belief by one group of creationists that the Earth is only 10,000 years old. Old earth age refers to the belief by another group of creationists that the Earth may be hundreds of millions of years old.

was the First Cause behind the origin of life." In other words, we didn't get here by chance alone.

Creationists maintain that evolutionists have tried to explain the origin of the universe by resorting to the same laws that describe its operation. "It's much like attempting to explain the origin of a TV set by the various laws that govern the operation of the TV," says Babu G. Ranganathan of the Institute of Creationist Research, a creationist advocacy group. The laws of chemistry and physics work well when they explain how living things function and operate, but they are inadequate when it comes to explaining the origin of the biological order or complexity in living things. In Ranganathan's view, "Evolution and creation are both different philosophical interpretations and conclusions of the scientific data." The issue is which philosophical position of origins is correct, creationists say.

How Could Life Come from Nothing?

Creationists assert that evolutionists run aground when they try to explain how something (life) could emerge from nothing (nonlife). Because spontaneous generation is ruled out, they ask, what other explanation is there? "Even in the laboratory, scientists with all their intelligence, planning, and sophisticated equipment and technology have not been able to create a single living cell from nonliving matter." Scientists may be able to do genetic engineering, but they are using only one form of life to manipulate another, not create new life from no life. And, creationists say, "Ultimately, however, life may be more than just creating the right chemical framework and structures because, even a dead cell, shortly before it decays, possess all of the proper and necessary chemical structures intact, but, alas, has no life."

Creationists also point to the fact that natural selection, as Darwin envisioned it, is limited to the genetic information already available in organisms. That is, natural selection can select for or against certain traits, but it can't act to promote or eliminate any trait if the genes that might produce that trait—by mutation, say—aren't there to begin with. For instance, if the genetic information for feathers and wings is not in the DNA of a lizard, that lizard will never develop feathers and wings. That causes creationists to doubt whether there is any scientific basis for believing that cumulative genetic mutations will produce the biological variation across different species that are demanded by Darwinian evolutionary theory.

Common Ancestors

When it comes to accounting for the origin of the human species, the creationists predictably take issue with evolutionists. "An interesting question that is often asked is how all the different races of humans could have descended from Adam and Eve," says Ranganathan of the Institute for Creation Research. That doesn't present a problem because we know that children with different characteristics can come from the same parents. "Adam and Eve, our first parents, possessed the genes to produce human offspring with different racial features or characteristics." But one thing that the creationists and evolutionists do agree on is that, because all humans are capable of breeding with one another, we are all descendants of a common human ancestor. The disagreement—and it's a huge one—comes in determining the identity of that ancestor.

Creationists also question the role of the fossil record in explaining evolution. They point to the absence of transitional links that would show one species evolving into another. "There are no fossils of partially-evolved species to indicate that there was a process of evolution going on," argues Ranganathan.

Fossil evidence used to support human evolution is also called into doubt. For one thing, even paleontologists and anthropologists give different interpretations of the same evidence. Moreover, many of the so-called "missing links"—those ape-men that evolutionists believe link prehumans to *Homo sapiens*—have turned out to be hoaxes (such as Piltdown Man) or nonhuman (such as Nebraska Man). Creationists assert that "there are no actual transitional links between ape-like forms and humans." Those remains that paleontologists have found, contends Ranganathan, "are mostly reconstructed from a few bones and most of what has been reconstructed is from imagination." Scientists may be too eager to find evidence of human evolution and so because "of the similarities between apes and humans it is very easy ... to read into partial and incomplete ape bones human qualities."

By the same token, creationists say, the similarities in the structures of different organisms—remember the link between whale flippers, bat wings, and human arms?—that supporters of evolution call upon to make their case do not prove evolution. These similarities do not suggest a common biological ancestry. According to Ranganathan, creationists believe that "the Creator designed similar functions for similar purposes and different functions for different purposes." In their view, neither the fossil record nor genetics support the idea of a common biological ancestry.

Ranganathan concludes his argument by underscoring the differences between creationism and Darwinian theory: "The Christian Scriptures teach that the world began in a perfect and harmonious state which became imperfect and disharmonious because of the sin of our first parents, Adam and Eve. This is the opposite direction of what the evolutionary theory teaches."

The creationists' basic position is based on the following principles:

➤ Science cannot investigate what is not observable. The origin of life is not observable, so, therefore, it cannot be studied.

➤ Evolution and creationism are two philosophical interpretations of the same scientific data: Evolution is not a science.

➤ There are no transitional links or partially formed species to connect species in rock formations, and thus there is no evidence of evolution from the fossil evidence.

➤ Similarity of physical or functional structures in different species, taken as evidence of common ancestry by evolutionists, can be accounted for by the design of a creator.

➤ No credible prehuman remains have ever been uncovered to establish a link between humans and apes (or apes to earlier species); those remains that have been found are likely to belong to some species, of ape. All human beings trace their ancestry to Adam and Eve.

Creationism Today: New Earth and Old Earth Ages

Creationism movements, bolstered by increased budgets, continue to push their agenda and expand their bases of support. According to *Scientist* magazine, the Young Earth Creationists (YEC) have two organizations with deep pockets: the Santee, California–based Institute for Creation Research, and the newer Answers In Genesis. YEC has established branches in Australia, Korea, Russia, and Turkey. In this country, at any rate, creationists would seem to have fertile ground in which to work. A 2001 Gallup poll suggests that more Americans—about 45 percent—believe in some form of creationism than they do any theory of evolution. Public opinion on the subject hasn't shifted significantly since Gallup first started asking about evolution and creationism in 1982.

Efforts by creationists to reshape school curricula met with their greatest success in Kansas in 1999, when the Kansas State Board of Education voted 8–4 to remove evolution as the sole explanation of the origin of humans. Henceforth, creationism was also to be taught as an alternative and equally valid theory. The decision raised an outcry from all over the country, particularly from scientists who expressed alarm that American schoolchildren would graduate without ever knowing anything about evolution. Their concerns were not unfounded: Many teachers, not wanting to provoke controversy, simply avoided the whole topic by not teaching either evolution or creationism.

In a Republican state primary, Kansas voters turned out three conservative members of the school board who had voted for adding creationism to the curricula. In 2001, when the new board met, it voted 7–3 to reinstate evolution as the new state science standard. An accompanying document, however, took account of the feelings of conservative opponents:

> While students may be required to understand some concepts that researchers use to conduct research and solve practical problems, they may accept or reject the scientific concepts presented. This applies particularly where students' and/or parents' beliefs may be at odds with the current scientific theories or concepts.

The board wanted to emphasize that just because a student was required to "understand" evolution, this was not to imply that belief in its veracity was also expected. Teachers were also urged to exercise discretion and not "ridicule, belittle, or embarrass a student for expressing an alternative view or belief." Instead, teachers were encouraged to "explain why the question (of creation) is outside the domain of natural science and encourage the student to discuss the question further with his or her family and other appropriate sources."

In spite of the inroads that creationists have made since the publication of *The Genesis Flood,* the 1967 Supreme Court ruling against teaching creationism in public schools caused many of its supporters to shift their focus. They sought to refashion creationism to give it more scientific respectability than YEC could hope to achieve. The result of their efforts is called intelligent design theory, or IDT, a controversial idea in its own right that we'll take a look at in our next chapter.

The Least You Need to Know

➤ Creationism is a system of belief that repudiates evolution as the basis for the origin and development of life on Earth (and the creation of the universe), asserting that all life owes its origin and existence to a First Cause, usually God.

➤ Before World War I, evangelical Protestant fundamentalists did not actively campaign against the teaching of evolution in public schools.

➤ Fundamentalists were convinced that World War I was waged as a way of realizing the social Darwinian idea of survival of the fittest.

➤ Several states, including Tennessee, passed laws banning the teaching of evolution in public schools in response to the fundamentalist movement led by William Jennings Bryan, a three-time presidential candidate.

➤ A Tennessee high school teacher named John Scopes broke the law to make a test case—the Scopes Trial (also known as the Monkey Trial) caused a great sensation when it was held in July 1925, although its outcome was inconclusive.

➤ The creationism movement has enjoyed a resurgence in recent decades, beginning with the publication of George MacCready Price's book *The Genesis Flood* in 1951.

Chapter 24

Building a Better Mousetrap: Intelligent Design

In This Chapter

➤ How intelligent design theory (IDT) differs from creationism

➤ The history of intelligent design

➤ The argument for intelligent design

➤ Opposition to IDT

➤ Is Darwinism incompatible with religion?

In the 2000 presidential race between George W. Bush and Al Gore, no one ever thought of asking either candidate his opinion about the validity of string theory as an explanation for the cosmos, even though it causes a lot of controversy and intense discussion among physicists. In fact, there was hardly any mention of science one way or another throughout the whole campaign.

But evolutionary theory was a notable exception. The hot-button issue was whether evolution should be taught alone or along with creationism. (For the record, Bush said that he favored teaching both.) Opponents of evolution question whether it is a "science" at all, asserting that it's simply a philosophical viewpoint. Politics and religion have seeped into the debate to such a degree that it's often hard to see what is really at stake.

Scientists and many public educators deplore what they perceive as a gradual erosion of academic standards, resulting from a politically driven need to accommodate supporters of creationism and intelligent design. It would be one thing if the public regarded evolution as a science in the same way that it does physics or chemistry, or even traditional biology. But because it is called a "theory" and is often misunderstood—and misrepresented—the Darwinian explanation for evolution has made for an easy target.

What's more, advocates of alternative explanations, such as creationism or intelligent design, have further muddied the waters by confusing evolutionary theory with evolution. Darwin's theory doesn't solve every problem, and it has a lot of holes, as its author was the first to admit. But that isn't the same thing as saying that evolution did not take place. While scientists clash over various aspects of the theory, most of them accept evolution as a natural phenomenon. Evolutionists question whether opponents of evolution are arguing from scientific or religious grounds. And if Darwin's foes were, in fact, arguing that there was an equally valid alternative to evolution based on science, how could they prove it?

Designer on Trial

For scientists, traditional creationism, with its roots in biblical literalism, hasn't been hard to demolish. Intelligent design theory, or IDT, is another matter. That's because adherents of IDT claim that theirs is a more scientifically viable theory than traditional creationism. They base their theory on "irreducible complexity," pointing out that many organic systems are so complicated and exquisitely organized—bacteria flagellum, for instance, or blood-clotting mechanisms—that all you have to do is remove one small part to wreak destruction. The same would hold true for a Swiss watch—take away a screw or a spring, and it won't tell time anymore. Given how intricate such systems are—and they are everywhere present in nature—then it follows that they must have been assembled by a cosmic designer, say IDT proponents. Is that hypothetical designer God? Most proponents of IDT would say "yes," although some might question whether the designer corresponds to the God of the Bible.

A Natural Selection

On April 23, 2001, the school board of Brodhead School District in Wisconsin voted 4–3 not to adopt two sets of science books that had been recommended by district staff after a year of review. One board member was quoted as concerned that the books presented evolution "pretty much as fact."

Getting Intelligent Design into the Classrooms

Advocates of IDT contend that because their theory is based on scientific evidence rather than biblical doctrine, it deserves to be taught as an alternative to

Darwinism. Unlike some creationists—notably, the young earth age advocates—intelligent design accepts the idea that the earth is billions of years old, not the thousands of years suggested by a literal interpretation of the Bible. Where they differ from Darwinians is in the idea that the action of natural selection alone can explain the complexity of the Earth's plants and animals.

Intelligent design was launched, just as modern creationism, by a book—in this case, Phillip Johnson's *Darwin on Trial,* published in 1991. In his book, Johnson, a professor of law at the University of California at Berkeley, put evolution on trial with the aim of persuading his readers that Darwinism is intellectually bankrupt. Johnson made no secret of his religious motivation. "We are taking an intuition most people have and making it a scientific and academic enterprise." By challenging Darwinism with IDT, he wrote, "We are removing the most important cultural roadblock to accepting the role of God as creator." While Johnson is no scientist, many theorists of intelligent design are reputable university scholars who accept evolution—up to a point.

Irreducible Complexity

IDT's leading advocate is Michael Behe, a biochemist, teaching at Lehigh University in Bethlehem, Pennsylvania. Behe, author of *Darwin's Black Box: The Biochemical Challenge to Evolution,* also worked for four years at the National Institutes of Health studying DNA. Although he is a devout Catholic, Behe shuns theological arguments, drawing instead on his knowledge of biochemical processes to assert that the observable complexity in living things can be explained only by the existence of an intelligent designer. Various biochemical structures in cells, he maintains, could not have been built in a Darwinian fashion, evolving gradually over time, because of an accumulation of favorable mutations. Behe's favorite example of an *irreducibly complex* mechanism is the simple mousetrap. It consists of at least five working parts—a platform, a spring, a hammer, a catch, and a holding bar—arranged in a specific way. If one part is missing, or if the arrangement is wrong, the mousetrap won't work.

Certainly the argument for intelligent design has an appeal that makes it seem, at least on the surface, like common sense. Complex order in nature

What Does It Mean?

A system *is* **irreducibly complex** when it consists of many interacting parts, all of which contribute to some function, so that the removal of any one part prevents the whole system from functioning.

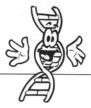

A Natural Selection

At the June 13, 2001, meeting of the school board of Plymouth Public Schools in Connecticut, two members recommended giving "equal time" to creation science or other alternatives to evolution in biology classes.

seems to cry out for some explanation other than the caprices of chance. Yes, say the IDT proponents, it would surely be a miracle if God had created Adam and Eve from the dust of the Earth, but would it be any less so if it turned out that God had created the genetic code that controls the development of human beings as well as all other living species? In other words, the IDT argument states that as long as the information-producing machinery owes its origin to a designer (or cosmic programmer), the designer is, therefore, responsible for the specific expressions of the information as well. In that light, you don't need to believe that Adam and Eve were the first human beings to see a designer's hand at work.

Poking Holes in Darwinism

The intelligent design movement has also made something of a name for itself searching for flaws in evolutionist thinking. Intelligent design has one advantage—the burden of proof falls on Darwinism. Put another way, Darwinism is guilty if it isn't proven innocent. And because Darwinian evolution hasn't been absolutely demonstrated, IDT proponents say that creation by an intelligent designer wins by default as the only reasonable alternative. At the same time, IDT defenders question the validity of Darwinism because they see it as promoting a materialistic view of the world that denies the reality of a spiritual realm. That suggests that IDT wants to have it both ways: Darwin is wrong on scientific grounds, but it is also wrong on philosophical and theological grounds.

A Natural Selection

In mid-2001, a Republican-sponsored "sense of the Senate" resolution stated that "where biological evolution is taught, the curriculum should help students to understand why this subject generates so much continuing controversy, and should prepare the students to be informed participants in public discussions regarding the subject."

Perhaps even more than creationism, proponents of intelligent design have had some success in getting their case across to the public. IDT is being given serious consideration as deserving "equal time" in school systems in Arizona, Ohio, and Minnesota. In a bid for respectability in higher academic circles, IDT has organized conferences at Yale and Baylor universities. In another campus victory for IDT, students at the University of California in San Diego have formed an organization called Intelligent Design and Evolution Awareness, or the IDEA club.

IDT advocates have also managed to get a hearing in state capitols. In Michigan, for instance, legislators have proposed putting intelligent design on an equal basis with evolution in the public school system. Similar moves are under way in Pennsylvania, where intelligent design theorists have teamed up with creationists. Lobbyists for IDT have focused their sights on Washington as well. In one instance, the Discovery Institute, an "IDT think tank" in Seattle, organized a briefing on intelligent design in 2000 for prominent members of Congress. Primarily funded by evangelical Christians, the Discovery Institute sponsors books—at least 25, to date—as well as conferences and fellowships for doctoral and postdoctoral research.

One of the Discovery Institute's major objectives is to win a case before the U.S. Supreme Court upholding the constitutionality of teaching IDT in public schools. That would be in marked contrast to the teaching of creation science, which the high court has ruled unconstitutional. The court, however, has indicated that teaching other *scientific* theories of origin is permissible, and it is that loophole that IDT is trying to use to slip into the classrooms.

IDT *is* science, asserts Steve Abrams, a Kansas school board member who originally voted for teaching creationism as well as evolution in schools (a decision rescinded in 2001). Abrams makes his claim on the grounds that IDT is based on "what is observable, measurable, testable, repeatable, falsifiable, good empirical science." The Discovery Institute argues that efforts to keep IDT out of the classroom while continuing to teach evolution represent "a victory for censorship and viewpoint discrimination." Just because it is a "dissenting scientific opinion on Darwinism," an institute spokesman declared, schoolchildren shouldn't be denied the right to learn about it.

IDT proponents don't feel very welcome by the scientific community, either. In a 1992 case, a biology professor at San Francisco State University was removed from his position by his chairman after criticizing Darwin's theories, although he was later reinstated. Certainly, scientists are concerned that IDT will make further gains at the expense of evolutionists. As a professor at Baylor University in Texas put it: "There's this feeling in the scientific community that this stuff needs to be shut down—that 'ID' is evil and if it succeeds it will overturn science."

A Natural Selection

In May 2001, a Minnesota appeals court turned down an appeal by Rodney LeVake who had argued that he had free exercise, free speech, and due process rights to teach "evidence against evolution."

Irreducible Complexity Does Not Compute

What is so wrong with the theory of intelligent design? Well, basically, most scientists believe that it is flat-out wrong. In spite of the claims of its advocates, intelligent design and irreducible complexity are not testable or falsifiable—you can't prove them one way or another, so IDT does not meet the criteria of science. "There's no dispute in science over *whether* evolution took place," said Eugenie Scott, executive director of the National Center for Science Education in Oakland, California, which promotes the teaching of evolution. "That's the big distinction lost in the general public." Scott dismisses the irreducible complexity concept because it fails to reflect an accurate understanding of natural selection. IDT uses a kind of rhetorical sleight of hand—a "God at the gaps" argument, as Scott terms it—in which God is called in to explain phenomena that are not yet well understood.

Filling the Gaps Without God

While creationists contend that evolution is also neither testable nor falsifiable, evolutionists counter that, quite the contrary, they do have the evidence on their side.

Why should it be necessary for God to fill the "gaps," they ask, when paleontologists are doing quite well on their own? For instance, in 1994, a skeleton of a "walking whale" (Ambulocetus) was found in northern Pakistan—an apparent intermediary organism between aquatic and land mammals. (Although the fossil record contains few transitional organisms, creationists are mistaken to claim that *none* have been discovered.)

Evolutionary geneticists are also helping to fill many of the gaps. *Mitochondrial DNA* evidence, for example, has lent credible support to the theory that all humans on the planet are descended from African ancestry. (Mitochondrial DNA is found in both the male and the female; only the female's mitochondrial DNA is transmitted to the offspring, though, while the male's mitochondrial DNA never penetrates the eggs. So, no genetic information is transferred. That makes mitochondrial DNA extremely valuable for tracing genealogical roots back many thousands of generations.)

What Does It Mean?

Mitochondrial DNA is contained on a chromosome found in both sperm and eggs, but only the eggs transmit the mitochondria DNA to the offspring. The sperm's mitochondria never penetrates the egg.

A Natural Selection

In Louisiana, a resolution opposing racism was introduced in the legislature in April 2001. It also asserted that Charles Darwin and his books promoted the justification of racism and that Adolf Hitler ultimately exploited these same views to justify killing millions of people.

Caught in His Own Trap

In one of the more dramatic confrontations between evolutionists and intelligent design theorists is Ken Miller, a Catholic biochemist at Brown University and a leading critic of intelligent design. Miller went face to face with IDT's leading advocate, Michael Behe, also a Catholic and a biochemist. As we mentioned earlier, Behe has often used the mousetrap as an example of irreducible complexity because it is a device that would presumably fail if any one of its five parts were removed. Miller demonstrated that the mousetrap could, in fact, be made to work even without one of its parts. Behe was unimpressed, asserting that Miller had only assumed the role of "designer" by reengineering the trap.

Behe says that he supports the teaching of evolution in schools but that he would like to see the problems with evolutionary theory discussed in the classroom as well: "I think it (evolution) should be taught warts and all." Behe adds that any discussion involving the origins of anything, whether life or the universe, necessarily involves philosophical implications. "People who think that teaching Darwinian biology does not touch on philosophical issues are kidding themselves."

Proving Evolution and Disproving a Designer

Evolutionists point out that supporters of creationism or IDT have some explaining of their own to do. For one thing, they say, if evolution doesn't occur, you have to come up with some way to account for how so many species have come into existence over time. Does that mean that the designer was compelled to miraculously intervene to separately create every species of life that exists today or in the past? That doesn't seem plausible. How did such an irreducibly complex mechanism in every living organism arise? If supernatural intervention does occur, exactly when, where, and how does it work? Were the first human beings created fully formed by a designer, and were they adults? Did they have bellybuttons? If so, why? Presumably, they wouldn't need bellybuttons if they'd never been born.

And while proponents of IDT deride evolution because they say that it cannot be observed, how can the actions of a hypothetical designer be observable? How does the designer execute these miraculous acts? Proponents of IDT tend to duck such questions, preferring the easier task of poking holes in evolutionary theory where "gaps" still exist.

Evolutionists call into question the very foundations of IDT. "What kind of scientific theory becomes established in our elementary, middle and high schools, rather than in the laboratory and through peer-reviewed journals?" writes Richard E. Lenser, a professor of microbioecology at Michigan State University. He then provides the answer: "None. No scientific theory can be validated by legislation demanding its inclusion in the classroom." Ken Miller, the Brown biochemist, makes much the same point. Advocates of intelligent design never publish their arguments in major scientific journals, he points out, mainly because they aren't interested in science so much as advancing a religious cause. "They are using political and social tools to gain acceptance in the classroom that they are unable or unwilling to win in the scientific community."

We should also point out that Darwin never claimed that his theory could be demonstrated so precisely or with such certitude that it would leave

Who Are They?

William Paley (1743–1805) was a British theologian and philosopher who wrote many popular works, among them *Natural Theology, or Evidences of the Existence and Attributes of the Deity,* published in 1863.

no room for reasonable doubt. No theory of the origin of species, Darwin admitted, could be proven absolutely. Some of the objections to evolution, he acknowledged in *The Origin of Species*, "are so serious that to this day I can hardly reflect on them without being in some degree staggered." And yet he insisted that his theory would emerge as highly "probable" to anyone who considered the "facts and arguments" in its favor. Darwin was acquainted with the kind of arguments that IDT proponents make. Attempts to introduce a "designer" into science date back to John Ray in the seventeenth century and later to the Rev. William Paley, author of *Natural Theology*, in the nineteenth century.

Can Science Talk About Creation?

"Most of our knowledge in science rests on probability rather than certainty," says Arthur Battson, a scientist and author. "If the alternative to Darwin's theory of evolution by natural selection is a theory of creation by a miracle-working designer, most of the empirical evidence is still on the side of Darwin." No one has any idea how to go about verifying the existence of a supernatural designer. But Battson isn't willing to tackle the question of creation, asserting that, by definition, creation is a "non-natural" event and cannot be investigated by science.

Some scientists agree that the study of evolution fails to provide evidence for or against "the operation of a superior being" in creation. Creationists, these scientists say, leave scientists no choice but "to continue to study the history of life as being controlled by materialistic mechanisms." In this view, Darwinism may not turn out to be "true," but it is the only theory that scientists have to work with. In other words, creationists may have a point—evolution may not be able to answer the question of how life got its start on Earth. But although Darwin's theory of evolution is unable to account for the origin of life, it can serve scientists very well in their study of the development of life on Earth.

Not every scientist agrees that creation is something that is out-of-bounds to science. "It is fine to scrutinize evolution," Battson says. Says another scientist, Richard Wakefield, "We have no fears of that—it is what makes us stronger." But he challenges creationists to clarify their understanding of how something was created out of nothing. "I don't buy the argument that it is supernatural and hence cannot be the subject of science. Hogwash!" Creation, he asserts, is a historical, not a supernatural, event, and scientists cannot arbitrarily put it up on the shelf and not deal with it.

Reconciling Darwinism and Religion

With all the furor surrounding this debate, you might think that modern science is incompatible with the moral, religious, and political traditions of America. That isn't true at all, though. In *Descent of Man*, published in 1871, Darwin expressed his belief in a natural moral sense rooted in humans. As social beings, he said, "[W]e need to cooperate with one another to succeed. Natural selection has favored those emotions—such as love, guilt and anger—that dispose us to cooperative relationships with relatives, friends and fellow group members." But we are also "intellectual animals," he observed, and so "we generalize our social emotions into the rules of good conduct and then into moral principles." Darwin then went on to emphasize that it isn't necessary to hold to any particular religious beliefs to have a moral sense, "but it shouldn't surprise us that religious teachings tend to support those universal standards of conduct—honoring parents, not stealing, refraining from unjustified killing and so on—that sustain social life."

Darwinism shouldn't be thought of as atheistic, although it has often been denounced as such by its detractors. Darwin allowed that scientific research could not answer questions about the first cause—the origins of life and the universe. That was where religious faith came in. As we noted, some modern scientists, such as Wakefield, are convinced that creation is a legitimate subject for scientific examination and that it shouldn't be considered a monopoly of religion alone. One way or another, the overwhelming majority of scientists consider Darwinian theory to be the best explanation for the origin and development of life on Earth that anyone has yet devised. From a scientific standpoint, creationism and intelligent design theory are expressions of belief and philosophical interpretations, but they do not meet the definition of science and so cannot be equated with evolution—nor do they deserve "equal time" in the public schools alongside evolution.

The Least You Need to Know

➤ The theory of intelligent design (IDT) states that Darwinian theory cannot account for all the "irreducible complexity" seen in living organisms and that only a cosmic "designer" could have been responsible for the natural order.

➤ Irreducible complexity, proposed by the IDT advocate Paul Behe, refers to the elaborate orchestration of all an organism's parts so that the removal of any one part will cause the organism to malfunction or die.

➤ Since the publication of *Darwin on Trial* in 1991, the IDT movement has had some notable success in getting its views aired in Washington and in state capitols and on college campuses.

➤ IDT advocates say that because evolution cannot be tested or observed, it has no right to claim to be a science.

➤ Although some scientists support IDT, the majority of scientists do not.

➤ Often lost in all the controversy about evolution is the recognition that nothing in Darwin's theory is incompatible with morality or the teachings of most major religions.

The Future of Evolution

What seems to disturb many people about evolution is not necessarily the mechanics of it—the way in which genes change from one generation to the next because of natural selection—but rather how evolution has produced such complex organisms without a blueprint. The idea that all of life on Earth could have come about by chance seems so unsettling and contrary to the instinctive need of humans to derive meaning from their lives. So, it isn't surprising that creationism and intelligent design theory carry so much appeal.

But even while battles still rage over the teaching of creationism and IDT in the schools, an even bigger storm is brewing just over the horizon. Now that we are poised at the beginning of a new century—indeed, a new millennium—we have it within our capacity to radically alter the course of evolution. Maybe nature didn't provide us with a blueprint, but, never being a species known for modesty, humans are now thinking seriously about drawing up one of their own. Of course, humans have tinkered with genetic inheritance all along—that's how we produced cows that give more milk, dogs that will obediently follow us to the ends of the Earth, and crops that resist pests. But never before have we possessed the means with which to change our own species, even to the extent of one day possibly defying death itself.

First of all, we need to address a question that has puzzled many students of evolution. Over the past half-billion years, no new phyla—such as vertebrates and plants—have arisen. Even fewer phyla exist now than in the past. Why this pause in evolution has occurred is unknown. "If you look back 500 million years ago, early history of invertebrates, there was an enormous range of designs which we no longer see on the Earth," notes the Harvard paleontologist Stephen Jay Gould.

In his view, the history of life doesn't show a progressive advancement of complexity. Evolution is more a screening process—sifting through the various designs available and choosing by natural selection the most promising forms. It's these successful variations tapped by natural selection that eventually developed into the relatively fewer highly successful forms that we see around us today. If humans have triumphed to become the dominant species on Earth we have nothing—and no one—to thank for it; it was simply the luck of the draw. "We are," he has written, "whatever our glories and accomplishments, a momentary cosmic accident that would never arise again if the tree of life could be replanted from seed and regrown under similar conditions."

Punctuated equilibrium, which Gould advocates, may also explain the apparent "hold" on evolution. According to this theory, species tend to go through long periods of "stasis," in which little or no evolution occurs, only to suddenly experience a burst of evolutionary change over a short period of time, geologically speaking.

But we mustn't forget that we are a very young species—50,000 years old, give or take—and so we wouldn't expect to see much perceptible evolution taking place among human populations. However, as we'll see, humans have acquired an unprecedented power to alter the course of evolution, by genetic manipulation and by the impact that we have on our environment. By taking matters in our own hands, we may actually accelerate evolution far beyond what would be possible if nature were left to her own devices. As Gould would say, that development may or may not represent progress.

The Biological Century

For all the furor it caused, the Darwin-Wallace theory of evolution by natural selection didn't have much immediate practical effect when it was introduced in the middle of the nineteenth century. Far more important to the lives of the people of Victorian England were the advances in electromagnetic technology by Michael Faraday and James Clerk Maxwell. By the end of the nineteenth century, the benefits of radio and electric light were only beginning to be appreciated; relativity, quantum mechanics, the Model T, and the invention of television and the Internet still lay in the future.

But even if people weren't able to grasp the full ramifications of the technological revolution they were living through, they understood that they were in the midst of great historical change. If one science was responsible for this change, it was physics, hands down. Think of radio, TV, the telephone, computers, jet planes, rocket ships, fiber optics, satellites, the internal combustion engine that gave us cars—and, less benignly, nuclear weapons. All those inventions came about principally because of advances in physics.

Like our forbears, we are poised at the brink of a new age, but the dominant science has changed. If physics dominated the twentieth century, biology is almost certain to dominate the twenty-first. And the promise—and the perils—of our biological century will certainly bring with it a radical shift in the way we think about life, evolution, and even the nature of what it means to be human, whether we like it or not.

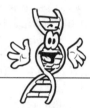

A Natural Selection

The Human Genome Project, carried out jointly by scientists working for an international consortium and a private biotechnology company, had two major objectives:

➤ To map all the genes in the nucleus of the human cell—about 30,000—and to establish where those genes are located on the chromosomes

➤ To sequence the genome by determining how genetic information is conveyed in the order of the DNA's chemical subunits

Unraveling the Mystery of Life: Sequencing the Human Genome

If there is one event that can be said to mark the beginning of the biological century, it's the sequencing of the human genome, completed in 2000, to date the largest biological effort ever undertaken. The genome that was mapped didn't belong to one person—it was actually a composite. Scientists had to take genetic samples from several people; that's because we are genetically the same, only different—about one percent different. Now that scientists have revealed the "map" of the human genome, they face the equally daunting task of filling in the blanks. Their objective is to reach a much deeper understanding of how genes influence a range of traits and behaviors.

Genes do not make us who we are—they have an influence over us, but their ultimate expression depends on our interaction with the environment. Nonetheless, unlocking the secrets of the genome will tell us a great deal more about how genes can make one person predisposed to a particular illness while conferring immunity to the same disease in another.

A Natural Selection

The sequencing of the entire human genome was completed on June 26, 2000, the result of an uneasy collaboration between the publicly funded Human Genome Project, an international consortium coordinated by the U.S. Dept. of Energy, and Celera Genomics, a private company in Rockville, Maryland.

What Does It Mean?

Genetic engineering refers to the alteration of an organism's hereditary material to produce desirable traits or eliminate negative ones. Genetic engineering is used for therapeutic purposes, to increase plant or animal food output, and to produce drugs or dispose of pollutants, among other uses.

The Temptation to Play God

All this astonishing new knowledge holds the promise of new drugs that may cure these hereditary diseases or keep them from happening at all. At the same time, the very capacity to identify gene-causing diseases raises a number of serious ethical issues. There was good reason why God didn't want Adam to eat of the fruit of knowledge. For instance, once we know who has defective genes, what will we do with that information? It's one thing for all people to be considered equal before the law, but, biologically, the fact is that we are not born equal. Some people have more of a genetic predisposition to illness than others, a fact that has not gone unnoticed by insurance companies when they decide whom they want to cover.

Well, you say, in the short term, it's quite possible that there will be abuses—that's why we're so concerned with guarding the privacy of medical records—but, sooner or later, won't scientists be able to fix the defective genes? The answer, guardedly, is that most of the genetic errors that have crept into our genome can be corrected, at least in theory. But what then? Once we know how to read our own genetic code, someone will think: Why stop there? Why not rewrite the genetic "text"? Why not play God?

Rewriting the Book of Life

The ability to tinker with, if not completely rewrite, the genetic code that influences what and who we are is already in our possession. True rewriting, as opposed to "editing" and "proofing," may not happen for some decades. But a quiet revolution has been going on for some time. This quiet revolution—though it's becoming noisier—is also based on *genetic engineering*. It's happening so routinely that, until recently, not many people outside of industry or the scientific community really noticed. Over the last several years, though, we have seen the development of genetically-engineered drugs, pest-resistant plants, single-gene alterations in plants and animals, and genetic diagnostics such as DNA "fingerprinting" used on criminal suspects.

Industrial applications are growing, too: For example, scientists have created genetically-engineered bacteria that obligingly devour up oil spills or other toxic pollutants. Predictably, some of these technologies have been applied to the marketplace, which explains why we have redder and riper-looking (and less tasty) tomatoes, specifically engineered to look more attractive on supermarket shelves. In a couple of decades, we are likely to see "bioactive" products that we might incorporate into our daily lives without thinking twice, including engineered microbes that clean our drains.

A Natural Selection

Plant–based vaccines hold special promise for Third World countries, where affordability, storage, and distribution of drugs are a problem. It may become possible for people in these areas to inoculate themselves against diseases by growing a crop of genetically-engineered fruits or vegetables, which they would need to consume only a few times a year.

Mimicking Evolution

Researchers are not waiting for evolution to happen; instead, they are using evolutionary principles to "breed" new proteins, refining specific traits—heat tolerance or the ability to stick to a cancer cell and isolate it—by tinkering with the underlying genetic code and selecting only the best of the new genes they are interested in. Essentially, they are mimicking the basic processes of evolution—producing genetic diversity and selecting desirable traits—to improve proteins. Ironically, these researchers are succeeding in many cases without even understanding the complex structures of the proteins they are working with. It's a perfect example of natural selection, except that there's nothing "natural" about it. For one thing, you can't count on natural selection acting to produce only the traits that you want to see—natural selection is seldom so accommodating. For another thing, natural selection takes time.

Why wait when you can speed up the process? Molecule by molecule, researchers are capable of coaxing a protein toward perfection in just a matter of weeks. The objective is to develop safer medications, more potent cleansers, and healthier crops. What works in the lab, though, doesn't always work in real life. A detergent enzyme for Procter and Gamble that performed well in lab tests, for example, failed when it was put to the test in an actual washing machine. All the same, this line of research underscores the fact that humans are already actively participating in evolution. The same techniques that can be made to select desirable traits in a protein can be used on humans—after all, we are all made of the same "stuff." But saying that is quite a bit different from declaring that we are masters of our own fate.

Researchers are also exploring ways of turning the traditional farm into "pharms," where the product being grown isn't wheat, say, but proteins. Scientists are able to synthesize proteins in cows, for example, by essentially turning their milk-making functions into protein manufacturers. Similarly, genetically-altered goats can produce proteins in their milk that can dissolve clots responsible for coronary blockages. Cows may one day be capable of producing insulin to treat diabetes, which, in all likelihood, will be much cheaper than insulin made in the labs. All it takes is just some careful editing of the relevant genes. But these enhanced animals would be unable to reproduce and pass along their insulin-making capability in their genes, so, while they may prove to be a rich source of pharmaceutical products, this kind of editing is unlikely to have any impact on evolution.

The Backlash

Not everything about this quiet revolution has been welcome. On the contrary, a storm of protest has erupted over genetically modified crops and led to bans against their importation in Europe. The backlash to genetic modification isn't limited to food, either. Throughout the early 1990s, the University of California at San Francisco was locked in a costly, ultimately unsuccessful battle with nearby residents who objected to virtually any genetic research that it carried out. The university was accused of contaminating the neighborhood with infectious pathogens, toxic wastes, and radioactivity. One outraged citizen went on television to say that "those people are bringing DNA into my neighborhood." Even though many of these protests are based on ignorance or emotion, they reflect an understandable fear that science is barreling ahead without taking into consideration the impact of their work on the public.

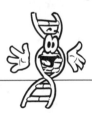

A Natural Selection

Using a growth hormone gene of the rainbow trout, scientists have used recombinant DNA techniques to create a form of carp that can grow to be one third larger than normal carp.

The Promise and Perils of the Biological Revolution

Where will this biological revolution—quiet or boisterous—ultimately lead? If the past is any guide, it won't be where we expect. No one ever said to himself, "Why not create a loyal pet that will love me even when the chips are down?" When some Paleolithic hunters tossed scraps of meat to curious wolves, it wasn't with the intention of creating a German shepherd or a cocker spaniel. And yet, by fostering a congenial environment for wolves with more docile natures, our ancestors unconsciously steered their evolution. In effect, they were engaged, without knowing it, in genetic engineering. Around 10,000 years ago, people began breeding and *hybridizing* plants and animals, selecting for the traits that they wanted in their grains and livestock. And while today we are far more aware of our evolutionary powers, much of the evolutionary change that we are bringing about is totally unplanned.

Although we usually associate evolution with long stretches of time, the evolutionary changes that humans are responsible for can happen surprisingly—and disconcertingly—fast. By overusing antibiotics, we have helped develop disease-resistant bacteria to propagate and spread. In many cases, our inadvertent use of natural selection has left us without any effective drugs to fight against new, ever more powerful and resistant strains. And antibiotics were introduced only in the 1940s.

Fishermen in the Pacific Northwest have also fallen victim to the law of unintended consequences. By overfishing pink salmon returning to the rivers of their birth to mate, they succeeded in reducing the population of big fish, encouraging the evolution of slow growth as a result. Pink salmon are now typically two-thirds the size of their ancestors that lived only 50 years ago.

By the liberal use of pesticides, farmers, too, have reshaped evolution, producing bugs that are resistant to increasingly toxic pesticides. To counter the damage they've done, scientists have inserted bacterial genes into plants to cause them to grow their own pesticides, but some experts fear that by helping to evolve resistant plants, scientists may be causing the evolution of mutant insects that can overcome the genetic resistance—and so on. Is there any surefire way of eliminating the threat of pests altogether, or will farmers have to make a trade-off and accept as inevitable the loss of some crops to insects? No matter how much editing and grafting of genes scientists perform, they will still

What Does It Mean?

Hybridization, also known as crossbreeding, involves breeding members of the same species with different characteristics or breeding members of different species, with the aim of combining desirable traits of both parents in the offspring. For 3,000 years, female horses have been bred with male donkeys to produce mules.

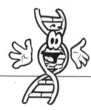

A Natural Selection

Selective breeding refers to selecting only plants and animals that have desirable traits for further breeding. For instance, corn has been selectively bred for its increased kernel size and number, as well as for nutritional value, for more than 7,000 years.

need to take into account the possibility that their research will have unforeseen, even disastrous consequences. But as our history shows us, for better or worse, once the genie (or genome) is out of the bottle, it's practically impossible to cram it back in. If something can be done, then, in all likelihood, it will be done.

Designing Babies: The Ultimate Shopping List

In India, upper-caste Brahmins use amniocentesis to determine the sex of a fetus early on in pregnancy; because sons are considered preferable, female fetuses are at risk of being aborted. In China, where sons are also favored and the government imposes restrictions on the number of children that a family can have to curb population growth, a similar phenomenon has been observed. As a result, China now has too many males reaching adolescence, and not enough females. (In China, there are now about 110 males to 100 females of reproductive age.) Nonetheless, the urge by parents to control whom they will bring into the world—and even what that child will look like—is simply too powerful to resist.

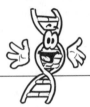

A Natural Selection

Recombinant DNA is also known as gene splicing. It is a technique in which scientists alter genetic material by introducing one or more genes of an organism into another organism. If the technique works, the genetic makeup of the recipient organism will be permanently altered. Recombinant DNA is a technique used in cloning.

Choosing Desirable Traits

It may take only a decade or two before it will be possible to screen fetuses for a vast range of attributes, such as the height they're likely to attain, what body type they will have, their hair and eye color, what sorts of illnesses they will be naturally resistant to, and even their IQ and temperament. And that's only the beginning: One day, at parents' instruction, doctors may be able to insert genes into fetuses that have been made in the lab to produce certain traits that wouldn't otherwise be possible. Some experts even see the day when parents can walk into a fertility clinic and pick out the traits that they want to see in their child from a list of options.

Needless to say, many people find this prospect alarming. "It's the ultimate shopping experience: designing your baby," says biotechnology critic Jeremy Rifkin, who, while repulsed by the idea, nonetheless recognizes that it might soon become a reality. "In a society used to cosmetic surgery and psychopharmacology, this is not a big step." Of course, any fad, especially one based on such an ephemeral attribute as attractiveness, is likely to run its course. If good looks or even intelligence can be acquired wholesale, these qualities may cease to be as valued as they are today. Scarcity, after all, plays a part in what society considers desirable, whether it's a brand-spanking-new Lexus or a supermodel's face.

Parents' quests for genes with the potential for creating beautiful children has had an interesting twist in that it has spurred a growing interest by academics into the possible evolutionary reasons behind certain physical characteristics: why, for instance, a certain hip-to-waist ratio attracts the eye or why symmetrical faces are considered desirable and beautiful.

The Coming Gen-Rich and Gen-Poor Divide

The same techniques applied to selecting desirable traits can be used to weed out undesirable traits. For example, doctors have developed a technique to identify fetuses with X-linked hydrocephalus, or water on the brain, which almost always affects boys. But what if the trait is only undesirable insofar as it's one the parents don't want for their child. What happens if they are told that their child will have only average intelligence or will be short or fat? Or what will parents do if they are told that their child can have certain "optional extras"—say, resistance to HIV? What about those children whose parents can't pay for either weeding out negative traits or inserting genes that will produce positive ones? Will we end up with yet another societal divide—the "gen-rich" and the "gen-poor"?

A Natural Selection

The use of donor eggs by infertile couples remains relatively uncommon. According to the National Infertility Association, about 1,700 babies were born from procedures involving egg donation in 1996. Those numbers have been growing only slightly since then.

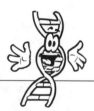

A Natural Selection

In 2001, the well-known infertility specialist Panayiotis Zavos, of the University of Kentucky, announced that he and Italian researcher Severino Antinori, who helped a 62-year-old woman give birth using donor eggs several years ago, were collaborating in an attempt to produce the first human clone. South Korean researchers claim to have already created a cloned human embryo, but say that they had destroyed it rather than implanting it in a surrogate mother to develop.

And what will that mean in evolutionary terms if such a development does occur? One thing that we are likely to see are exclusive dating services that try to match up people with "superior" genetic traits. After all, why should a woman who has been the beneficiary of a carefully tailored genetic endowment chosen by her parents accept as the potential father of her children any man who has not received a similar genetic legacy from his parents?

On the other hand, genetic manipulation is still a relatively new science, so it isn't going to be so easy to insert or remove traits at will. Nor does it follow that their efforts will pan out the way that the prospective child's parents envision. A baby may be carefully designed to be a wiz at math, a champion on the ball field, and an irresistible beauty to the opposite sex, and still turn out to be a disappointment.

Altering the Environment/Altering Ourselves

It's hardly any secret that humans have been altering their environment for many years, although lately that process has been accelerating: Global warming, caused by carbon dioxide emissions, is only the best-known example. What is less well known is the fact that the changing environmental conditions have changed us in ways large and small and, in doing so, have played a role in our evolution.

Recent studies out of Germany, for example, suggest that by changing our environment, we might already have changed our biology. In the last 20 years alone, the German investigators have been testing 4,000 subjects to measure the acuity of their senses of smell, taste, touch, sight, and hearing. Their findings are disturbing: On average, each of these senses has decreased at a rate of nearly 1 percent a year. "Fifteen years ago, Germans could distinguish 300,000 sounds," said the chief researcher, Dr. Henner Ertel. "Today, on average, they only make it to 180,000. Many children stagnate at 100,000. That is enough for hip hop and rap music, but it is insufficient for the subtleties of a classical symphony."

A Natural Selection

Through artificial insemination, women have long been able to have children without a male sexual partner. Soon it may be men's turn; using cloning techniques, a man could potentially become a father alone, replacing the nucleus of a donor egg with his own and then finding a woman to carry the child to term.

Related studies have shown "generation gaps" between brains formed before 1948, brains formed between 1948–1968, and brains formed since. According to scientists, the more recently the brain was formed, the more "dissonance" it can tolerate. Dissonance is "noise"—confusing stimuli that brains usually tune out, so as not to become overwhelmed by irrelevant distraction. But, say the scientists, probably as a result of environmental influences, the brains of people born after 1968 are able to accommodate floods of contradictory information without blotting it out.

The researchers didn't see this as a positive development at all; instead, it suggests to them that the filtering processes observed in older brains are not working as well in newer brains, causing a diminishment of awareness. "We are seeing the largest and fastest breakthrough since the dawn of consciousness," declared Ertel. "Our brain is not adapting. It is rebelling against the world and changing it by changing itself. Red is no longer red. Sweet smells begin to stink. In the next century, different people will be living in a new world."

It's difficult to say, on the basis of a few studies, whether Ertel's dire predictions will turn out to come true. We still haven't gathered enough data to say with certainty that the environment is having as much of an evolutionary impact as these German studies suggest or whether, even when changes do occur, they are very widespread in the human population. But there is no dispute that we are re-creating the environment in very different ways—from creating "pharm" animals and vaccine-producing plants to polluting oceans and raising the temperature of the planet. As we do so, we are unconsciously and inadvertently setting in motion evolutionary processes that may change the human species as well.

The Immortality Game: Defeating Age and Death

Humans are also flirting with an even more radical idea than simply altering the genetic makeup of a person: stretching the natural envelope by eliminating aging and death altogether. In the biologist's view, it is differentiation of cells that makes us mortal. No organism of any complexity would be able to develop if the cells from which it is formed didn't begin to divide and differentiate. (Otherwise, we would remain simple unicellular organisms.) These cells turn into our organs and tissues, becoming brain, heart, pancreatic, and muscle cells. Once these cells have reached maturity, however, they no longer divide at the same rate. Skin and blood cells can regenerate quickly, while heart and brain cells do so more slowly. (Until the late 1990s, scientists didn't believe that brain or heart cells could regenerate at all.) Nonetheless, these adult cells can regenerate only up to a point. The same restriction doesn't apply to embryonic stem cells, though.

In the embryo, a small number of stem cells are set aside before embryonic development begins. Throughout the embryo's development, these privileged cells are protected from differentiation—essentially, these cells are immortal. Most diseases involve the death of healthy cells—for example, brain cells in Alzheimer's disease, cardiac cells in heart disease, and pancreatic cells in diabetes. Once scientists discovered that embryonic stem cells have the potential to develop into any cell that is

What Does It Mean?

Pluripotent is a term meaning not fixed in developmental potential. It is usually applied to embryonic stem cells, which, because they are so plastic and adaptable, have the capacity to develop into any organ or tissue. The term was first used in 1916.

needed—brain, heart, liver, arms, hands, retinal skin, and nerve cells—they realized that they could use the body's own repair kit in treating diseases that had long eluded a cure. The embryonic stem cells, which have a pedigree stretching back more than three billion years to the first cell, are called *pluripotent* because they can take up residence anywhere and turn into the tissue or organ as needed.

The Search for Immortality

Although some wealthy individuals are putting their bodies (or, for less cost, their heads alone) on ice in hope that science will ultimately be able to revive them, others hoping for immortality don't feel that they have to wait that long. They are counting on techniques developed from embryonic stem cells to extend their lives well beyond the longest life span conceivable now. (There is no record of anyone living much beyond 120.) About a decade ago, scientists discovered that each time cells replicate (reproduce themselves), the tips of their chromosomes known as *telomeres,* shorten until the process reaches a point at which the cells simply stop dividing altogether.

This point, called the Hayflick limit, comes after about 50 replications. Some scientists believe that the continual shortening of the telomeres is responsible for aging. Biochemists have succeeded in reactivating the enzyme that lengthens the telomeres. In lab cultures, they have extended the lives of cells by at least 20 divisions past the Hayflick limit. By reconstituting the telomeres of embryonic stem cells, scientists hope that it will become possible to rejuvenate parts of any organ with a simple injection. However, there is some question of what role telomeres actually play in aging; in practical terms, no one can be sure that extending their capacity to divide will ultimately lead to longer life spans in humans.

What Does It Mean?

Telomeres serve as a kind of cap on the ends of chromosomes to keep them from attaching to other chromosomes. Some scientists believe that the shortening of the telomeres that occurs with each cell division may play an important role in the aging process.

The objective of the scientists involved with this research is to ultimately use telomere-lengthening techniques—assuming that they can prevent aging—on all the body's tissues, continuously regenerating them from permanently youthful tissue stem cells descended from embryo stem cells. That would allow individuals to regenerate their organs and tissues when their own either run out due to wear and tear or are affected by disease.

The Potential of Cloning: Double or Nothing

Alternatively, some scientists believe that clones could be developed to supply the organs and tissues that people need as replacements when the time came. "The miracle of cloning isn't what people think it is," explained a biochemist working in the field. "Cloning allows you to make a genetically identical copy of an animal, yes, but in the eyes of a biologist, the real miracle is seeing a skin cell being put back into the egg cell … which then can turn into any cell in the body." This means that *cloning* isn't necessarily just about producing copies, but it also can be applied to generate new tissue to replace injured or diseased parts of the body.

What Does It Mean?

Cloning is a form of recombinant DNA in which scientists remove the DNA-containing nucleus from the egg of a female and replace it with a nucleus from another animal of the same species. The egg is then planted back into the uterus, where it develops into a "copy" of the animal from which the transplanted nucleus was taken. The technique, still in its infancy, was used to produce the cloned sheep Dolly in 1997.

Even given this spin, there's no question that cloning arouses a lot of emotion—perhaps more than it should, given its rarity and the difficulty of actually producing a clone. Supporters see nothing wrong with it, regarding the technique as simply another reproductive tool. If parents want to reproduce their children, whether alive or dead, they say, why should they be stopped? But theirs is a minority view. Most people are appalled by the practice, at least when it involves humans, and legislation has been introduced in several states as well as in other nations to ban it.

In theory, cloning isn't very hard. An egg is removed from a female, and its DNA-nucleus is extracted and replaced with a DNA-nucleus of a cell from an animal of the same species. Then the egg with the transplanted nucleus is placed back in the uterus of the recipient female; if it is carried to term, the offspring will be an identical copy of the donor animal. It doesn't matter—in theory—from which cell the nucleus comes because every cell contains all the organism's genes. In practice, however, cloning is very difficult to pull off. The most notable success is the cloning of the sheep Dolly by Scottish researchers in 1997. But Dolly came after hundreds of failed experiments, and while other species such as mice and cows have been cloned, scientists admit that they have a lot of kinks to work out before they can produce clones reliably.

The Limitations of Cloning

So, it's problematic but by no means impossible for a human clone to be produced. Again there may be unforeseen consequences once it happens. Some evidence has emerged to indicate that the process of cloning may impair the clone from being able to reproduce. In 1999, scientists associated with the Roslin Institute, where Dolly was created, reported that their prized sheep's DNA was aging faster than that of an ordinary sheep, which might leave her vulnerable to cancer and maybe an early death. Possibly her genetic mother's DNA—derived from a mature udder cell—might run up against the limit of an absolute age. If such a limit does exist and the limit applies to most species, cloning might lose a good deal of its allure (and the fear that it stirs in people). Rather than ensuring a form of immortality, cloning might turn out to be one of the quickest ways to wear out a bloodline.

Questions of morality and individual identity aside, cloning poses yet another risk where evolution is concerned: the possibility of undermining our biological diversity. That's because clones have identical DNA to their parent—there is no exchange of genetic information, usually achieved in mammals by sexual reproduction. And it's that diversity that accounts for why natural selection works at all.

Is the World Ready for Ultrahumans?

Some scientists are even thinking farther ahead, beyond embryo stem cells and cloning, envisioning a day when people could transform themselves into "ultrahumans," the living sum of their replacement (and added-on) parts. These future humans might have metabrain processors and emotion modulators (to better control their tempers), as well as supraorganic perceptual abilities, with bodies integrated with noncarbon compounds. And, of course, they would live forever, their immortality guaranteed by innovative genetic, cellular, or synthetic techniques. Should such an "upgraded" human being become feasible, no doubt marketing directors would have a field day, promoting "made-for-success" bodies. Why endure bodily, mental, or emotional suffering that has been the bane of humans since they first emerged in Africa, when they could be freed from worrying about their bodies or being subject to capricious emotions?

Given the precedent, people will probably demand equal access to bodily "upgrades and endowments." But even if upgraded humans could live forever, they still wouldn't be able to have children with these same "immortal" traits because individuals can't pass on attributes acquired in their lifetimes. So, in that sense, an ultrahuman would represent an evolutionary dead end.

The Least You Need to Know

➤ Evidence suggests that evolution is continuing at a slower rate.

➤ Humans are now becoming responsible, to some degree, in the evolutionary process.

➤ In a multitude of industries, scientists are engineering plants and animals to increase output, turning farms into "pharms" for the production of proteins, vaccines, and other drugs, and creating bacteria to consume pollutants and industrial wastes.

➤ The prospect that the parents will be able to create children who are "preplanned"—designer babies endowed with high intelligence and attractive features and bodies—raises the specter of a social divide in which a small elite is "gen-rich" while a larger number of people, unable to afford to design their babies, will have to resign themselves to being "gen-poor."

➤ The search for immortality has never slackened, but now, thanks to science, the prospect exists that one day humans will be able to keep themselves alive well beyond the limits of human life spans today by a combination of genetics, lab-grown body parts, and synthetic components.

Glossary

abiotic factors   Physical factors in an organism's environment such as water, climate, light, and oxygen.

albinism   A genetic abnormality that depletes the skin of pigmentation.

alkaptonuria   A hereditary ailment that causes urine to turn a deep reddish brown.

alleles   Different forms of the same gene contributed to the offspring from each parent.

allopatric speciation   Creation of new species by the physical isolation of a population. *See* speciation.

analogous structure   Body structures that appear similar in different species.

artificial selection   Breeding or hybridization of plants and animals by humans to produce desirable traits in the offspring.

athenosphere   A layer of Earth made up of rock supporting the rigid tectonic plates that make up the outer surface of the Earth. *See* plate tectonics.

Australopithecus   A hominid that lived about four million years ago; its name means "southern ape," after its discovery in South Africa. Three separate types of Australopithecus have been identified: *Australopithecus africanus, Australopithecus robustus,* and *Australopithecus afarensis.*

bases or **base pairs**   Nitrogen-bearing compounds that make up the structures of DNA: adenine, guanine, thymine, and cytosine.

bestiaries   Medieval books filled with descriptions of all the animals in creation, including imaginary ones such as the unicorn and the phoenix.

binomial   Literally, "two names"; a naming system for plants and animals, in which the first name refers to the genus and the second refers to the species.

biodiversity   The total number of different species of animals, plants, fungi, and microbiotic life on Earth, and the variety of their habitats.

biogenetic law   A discredited theory proposed by nineteenth-century German biologist Ernst Haeckel that sought to assign a ranking to different species from the lowest to the highest.

biological evolution   Descent with modification; any change in the properties of populations of organisms that transcend the lifetime of a single individual.

biotic factors   Influences on an organism from other living organisms such as predators, parasites, and prey.

bipedalism   The ability to walk upright on two feet.

Cambrian Explosion   A burst of evolutionary activity 540 million years ago that produced vast numbers of new species.

carrying capacity   The limit to a population's growth because of environmental or other stresses.

catastrophism   A theory popular in the eighteenth and early nineteenth centuries that seeks to explain the present state of the Earth's surface as arising from a series of violent geological cataclysms, particularly the Biblical Flood.

Cenzoic Epoch   A geological period from 65 million to 1.6 million years ago.

chromosomes   Microscopic units in each cell containing DNA.

chronomatic measurement   A nineteenth-century method of assigning a calendar year date to artifacts, fossils, and other remains.

clade   A group such as a species, whose members all share a common ancestor.

class   A classification of organisms with more characteristics in common than in either a kingdom or a phylum. Apes, monkeys, and mice are all in the class Mammalia.

continental drift   A theory proposed by German scientist Alfred Wegener stating that all present-day continents once composed a single land mass that began to break apart in successive stages and drift away from each other, eventually reaching their current locations.

creationism, also **scientific creationism**   A belief that the origin of life on Earth and all existing species is attributable to some form of divine intervention.

Cro-Magnon   Early *Homo sapiens,* who inhabited western and southern Europe during the last ice age about 120,000 years ago.

cryptic   A protective adaptive strategy in which organisms camouflage themselves to blend in with their environment.

day-age theory   A creationist idea that the "days" of creation were metaphorical, suggesting that creation might have taken hundreds of millions of years.

direct fitness   A measure of how many alleles, on average, a genotype contributes to the next generation's gene pool by reproducing.

DNA (deoxyribonucleic acid)   The genetic material in the cells of all organisms and viruses that contains hereditary information.

dominant allele   A gene form from one or both parents that codes for a trait that will express itself in the next generation. *See* recessive allele.

ecological niche   A particular habitat to which a species or population has adapted.

ecosystem   A system made up of an environment and the organisms that inhabit it.

empiricism   Knowledge acquired from experience.

entomology   The study of insect life.

enzyme   A specialized organic substance composed of amino acids that acts as catalysts to regulate the speed of chemical reactions in organisms.

Eocene Epoch   A geological period from 55 million to 38 million yeas ago.

epigamic sexual selection   Mating ritual in some species, usually carried on by the females.

ethology   The study of animal behavior.

eugenics   A doctrine born in the nineteenth century that claims that a society's heredity can be controlled by regulating who has the right to reproduce.

extinctions, background rate   The average rate of extinction since the last mass extinction 65 million years ago, used to gauge the current rate of extinction.

extinctions, mass   The elimination of a great proportion of the planet's species by an environmental catastrophe; five mass extinctions have been recorded in the 4.4-billion-year history of Earth.

family   A classification to describe organisms with more characteristics in common than in a class. Wolves and cats belong to the class Carnivora.

fission   Asexual reproduction in which an organism divides into two or more parts, each of which develops into a complete organism.

fitness   The ability of an individual to pass more of his genes to the next generation.

formalism   A belief that the form or structure of an organism is more important than the function of any of its parts in determining the organism's behavior and habits.

founder effect   Diminished genetic diversity that may occur when a species whose original population has been nearly wiped out by some environmental disaster tries to reestablish itself in a new environment.

functionalism   A belief that function is more important than form in an organism and that modification of any one part will impair the function of the whole being.

gamete   The sexual reproductive cell that fuses with another sexual cell in the process of fertilization.

gene   A unit of inheritance, composed of genetic material.

gene flow   The introduction of new gene types into a population's gene pool through migration of the same or closely related species.

gene pool   The sum of genotypes in a given population.

gene recombination   A shuffling of genes by the exchange of genetic information between chromosomes; the recombined information can then be passed to offspring that will express the new traits.

genetic death   The destruction of certain genetic variants in a population.

genetic drift   A random process that adds alleles to or subtracts alleles from a population, usually resulting in a loss of genetic diversity.

genetic engineering   The alteration of an organism's hereditary material to produce desirable traits or eliminate negative ones.

genotype or **genetic variant**   The underlying genetic makeup of a trait, or the overall genetic makeup of the individual.

genus   A category that includes a group of species closely related in structure and evolutionary origin. Genus falls below family (or subfamily) in the classification of organisms.

Gondwanaland   The name given by Eduard Suess, an Austrian geologist, to a southern "supercontinent" composed of India, Africa, and Madagascar that was supposed to exist in the distant geological past.

good genes model   A theory to explain sexual selection that proposes that a male's extravagant display and bravado is an indication of fitness or capacity for reproductive success.

gradualism   A concept of Darwinian theory stating that new species are created through the gradual accumulation of many small genetic changes over long periods of time.

Hardy-Weinberg rule   A set of algebraic formulas describing how the proportion of different genes can remain the same in a large population over time.

Hayflick limit   The point at which cells stop dividing and replicating.

Holocene Epoch or **Post-glacial Epoch**   A geological period that took place about 10,000 years ago.

hominid   A family of prehumans and other creatures that originated in Africa three million to four million years ago, with the capacity to walk upright.

Homo erectus   A hominid that lived about one million to two million years ago and that was characterized by its ability to walk upright and make tools.

Homo habilis   A tool-making hominid that lived about two million years ago in east Africa. The name translates as "handy man."

Homo sapiens   The human species, which originated in Africa about 50,000 years ago—literally "knowing man."

homologous structures   Physical structures in different species that, although different in appearance, have the same underlying structure.

Human Genome Project   An international public and private effort to sequence the human genome, completed in 2000.

Huntington's disease   A hereditary brain disorder.

hybridization, also **crossbreeding**   Breeding members of the same species with different characteristics, or breeding members of different species to produce desirable traits of both parents in the offspring.

ice ages   Any of several periods, lasting millions of years, during which much of the Earth's surface was covered by ice and the planet's climate was marked by significant cooling.

ichthyology   A study of fish life.

incumbency   The evolutionary advantage enjoyed by a species that is first to occupy and dominate a given ecological niche.

index fossils   Fossils formed from species that existed only for a short time and that are particularly valuable in determining a rock's age.

indirect fitness   An individual's ability to pass along genotypes identical or related to his own by helping relatives to produce offspring.

intelligent design theory (IDT)   A belief that because life is so complex, it couldn't have come into existence as a result of chance but that this complexity instead shows evidence of a cosmic or intelligent designer, which may or may not correspond to the God of the Bible.

intermediaries   Transitional species that form a link between extinct and existing species.

intrasexual   A form of sexual selection in which males (and rarely females) compete through a display or a physical context for mates.

invertebrates   Any animal lacking a vertebral column, or backbone, a classification that includes the majority of the animal kingdom.

irreducibly complex   A belief of intelligent design theory stating that organisms are so exquisitely formed that the removal of any one part will cause the organism to stop functioning.

ladder of being   A concept first proposed by the Greek philosopher Aristotle assigning a hierarchical rank to each species. In the Middle Ages a similar system was introduced, known as the great chain of being.

Kenyanthropus platyops   A hominid believed to have lived in east Africa about three and a half million years ago.

kingdom   The highest and most general classification of organic life.

***laissez-faire* capitalism**   An economic doctrine that says that the best economic system is one with no interference by government.

Lamarckism   The idea proposed by the French naturalist Jean-Bapiste Lamarck that traits acquired by an individual during his lifetime could be passed along to the offspring.

lineage   A line of descent from a common ancestor.

macroevolution   Any evolutionary change at or above the level of species.

meiosis   A process of cell division, during which the cell's genetic information, contained in the chromosomes, is mixed and divided into sex cells.

Mendelian population   Any species in which all its members can reproduce only with one another and with no other species.

Mesolithic Era (Middle Stone Age)   An era about 13,000 years ago.

metamorphosis   A process in which insects such s butterflies begin life as larvae and undergo a dramatic change to another form.

microevolution   Evolution that occurs below the level of the species.

mimicry   A protective adaptive strategy in which organisms simulate traits of another species to deceive natural predators.

Miocene Epoch   A geological period that took place between 24 million and 5 million years ago.

mitochondrial DNA   A form of DNA contained on a chromosome that is found in both sperm and eggs but that is passed to the offspring only by the female.

modern synthesis   An effort to reconcile Darwin's theory of evolution based on natural selection with macroevolutionary theory based on modern genetic discoveries.

molecular electrophoresis   A method of studying DNA by separating and purifying DNA or proteins in an electrical field.

Monkey Trial   The famous 1925 trial of high-school teacher John Scopes, charged with violating a Tennessee statute against the teaching of evolution.

morphology   The study of structure or systems of a body such as muscles or bones.

mutation   Any alteration of the genetic information of an organism.

natural selection   The mechanism of evolution by which the environment acts on populations to enhance the adaptive ability and reproductive success of individuals possessing desirable genetic variants, increasing the chance that those beneficial traits will predominate in succeeding generations.

Neanderthals   A population of extinct hominids, whose lineage extends back from 24,000 to 200,000 years ago. Anthropologists disagree on whether Neanderthals are *Homo sapiens* or a separate species.

Nebraska Man   Fossil remains supposedly of an extinct prehuman found in Nebraska in 1927, later proven to belong to an extinct pig.

negative selection   The process of weeding out harmful alleles from a gene pool by natural selection.

Neolithic Era (New Stone Age)   A period that began about 10,000 years ago with the spread of agricultural villages through the Middle East and Central and South America.

nucleotides   Chemical compounds that make up DNA.

old earth age   The belief of some creationists that the Earth may be hundreds of millions of years old.

Oligocene Epoch   A geological period that took place between 38 million and 24 million years ago.

"Ontogeny recapitulates phylogeny"   A statement attributed to the German biologist Ernst Haeckel, meaning that the development of the human embryo retraces the evolutionary development of the human race.

order   A classification of organisms in which members share more characteristics in common than members of a class. Dogs and raccoons belong to the class Carnivora.

organism   A living being; an individual made up of interdependent organs.

ornithology   The study of bird life.

paleoanthropology   A profession combining paleontology and anthropology.

Paleocene Epoch   The geological period that took place between 65 million and 55 million years ago.

Paleolithic (Old Stone Age)   A period that began about two million years ago with the development of stone tools by the first prehumans and that ended about 13,000 years ago.

paleomagnetism   The preservation of magnetic fields in rocks.

paleontology   The study of fossils.

Pangaea   The name given by the Austrian geologist Alfred Wegener to the supercontinent that existed in the distant geological past before the fragmentation that led to the present-day continents.

panmictic unit   Any population in which random breeding occurs.

parthogenesis   The ability of certain plants and insects to reproduce without sex.

Peking Man   Fossil remains of a hominid discovered near Peking, China, in 1929 that disappeared at the beginning of World War II.

Permian Period   The last division of the Paleozoic Era that occurred between 290 million and 240 million years ago, characterized by dramatic geological upheaval.

phenetic evolution   Evolutionary changes that occur in response to environmental factors.

phenotype   The appearance of an individual produced by the interaction of genes and environment.

phylectic change, also **vertical change**   Gradual evolutionary change.

phylum (pl.: phyla)   A classification of organisms with a common design or organization. (*Phyla* refer to animals; the term *division* is applied to plants, fungi, and protista.)

Piltdown Man   Supposed fossil remains of an extinct prehuman found in Piltdown, England, in 1912, later proved to be a hoax.

Pithecanthropus   Ape man, referring to early hominids.

plate tectonics   A unifying theory of geology developed in the 1960s that builds on the theory of continental drift, based on the idea that the Earth's surface is composed of floating plates.

Platyops   Flat-faced, referring to one type of early hominid.

Pleistocene Epoch   A geological period that took place between 1.6 million and 10,000 years ago.

Pliocene Epoch   A geological period that took place 5 million to 1.6 million years ago.

pluripotent   Not fixed in developmental potential. The term is often applied to embryonic stem cells, which, because they are so plastic and adaptable, have the capacity to develop into any organ or tissue.

polygamy   The monopolization of members of the opposite sex by an individual for reproductive purposes.

population bottleneck   A dramatic change of environment, such as disease or famine, that reduces a population's size and genetic diversity.

primatology   The study of primate life.

principle of independent assortment   A Mendelian law describing how genes randomly mate and transmit genetic information.

principle of segregation   A Mendelian law describing the mechanism of inheritance by which both parents individually make genetic contributions to their offspring.

principles of faunal succession   Basic laws establishing the relationship of fossils and rock strata for dating purposes.

proteins   Organic compounds essential to the functioning of living organisms.

protist   A group of simple organisms with characteristics of both plants and animals.

punctuated equilibrium theory   The idea that evolutionary change occurs mainly in a spurt of evolutionary activity within a relatively short geological period, followed by longer periods in which little or no change occurs.

quadripedalism   Movement of an organism using all four limbs.

radiata   A classification of animals that have symmetrical bodies around a central point, such as jellyfish.

radiocarbon dating   A method of dating the approximate ages of extinct organisms, artifacts, rock formations, and so on by measuring the amount of carbon-14 that they contain.

recessive allele   A form of a gene that can come from either parent but that will express itself in the offspring only if recessive alleles for the same trait from both parents are inherited.

reciprocal altruism   Mutually cooperative behavior among different members of a species that enhances the chance of reproductive success of both donor and recipient.

recombinant DNA or **gene splicing**   A technique in which scientists alter genetic material by introducing one or more genes of an organism into another organism.

reproductive success   The advantage conferred on individuals with genetic variants that make them better able to adapt to new environments, reproduce, and pass along their genes to offspring.

runaway sexual selection model   A theory to explain sexual selection that says that females may have an innate preference for some male trait even before it appears in a population.

sagital   A ridge of bone that runs from the top to the bottom of the skull, found in species of ape and hominids.

savanna hypothesis   An explanation for the evolution of hominids and their split from apes eight million to five million years ago based on changes to the environment that created more savannas in Africa.

scientific creationism   *See* creationism.

Scopes Trial   *See* Monkey Trial.

selective breeding   A process of selecting only plants and animals that have desirable traits for further breeding.

sexual dimorphism   The occurrence of two forms of the male in the same species.

sickle cell anemia   An inherited blood disease caused by two recessive alleles, frequently seen in people of African descent.

social Darwinism   A nineteenth-century theory, based on the idea of a continual struggle for survival, that proposes that individuals, groups and races are subject to natural selection.

sociobiology   A developing field of science that investigates behavioral patterns among different groups, races, and populations of animals (including humans), in the belief that many types of behavior can be inherited.

speciation   The creation of a new species when one group in a population can no longer reproduce with the original group.

species   A distinct kind of organism, with a characteristic shape, size, behavior, and habitat, composing a population in which individuals can interbreed only with one another.

species selection   A macroevolutionary theory that says that natural selection favors species with favorable traits, just as below the level of species, it favors individuals with favorable traits.

spontaneous generation   The discredited belief that life could arise spontaneously out of inorganic material.

stratigraphy   The study of the layers of the Earth's surface.

survival of the fittest   The idea promoted by social Darwinists that, because of favorable traits, some individuals are better able to survive and reproduce under environmental pressures.

taxa   A group in a classification system.

taxonomy   A system of classification often applied to the classification of plants and animals according to their presumed natural relationships.

telomeres   Tips of chromosomes.

Turkana Boy   Fossil remains of a hominid of about 12 years old who may have lived about 1.6 million years ago, and classified as *Homo erectus*.

uniformitarianism   A theory stating that the natural processes that change the Earth in the present have operated in the past at the same gradual rate.

variability hypothesis   An explanation of why early hominids evolved, based on meeting the challenge of a wide range of environmental and climactic conditions in Africa.

vertebrates   A classification of animals, including mammals, that all have a segmented spinal cord.

Woodland-Mosaic hypothesis   An explanation for the evolution of hominids based on their ability to adapt to terrain in Africa that was partly woodland and partly grassland.

young earth creationism   A belief based on a literal reading of the Bible that the Earth is only about 10,000 years old.

zygote   The cell resulting from the fusion of a male and a female sex cell that then must undergo many cell divisions before it develops into an organism.

Timeline

15–12 billion years	Big Bang: Birth of the universe
4.6 billion years	Formation of the Earth
3.8 billion years	Earth's crust solidifies—formation of oldest rocks found on Earth. Condensation of atmospheric water into oceans.
3.5–2.8 billion years	Simple prokaryotic cell organisms develop. Beginning of photosynthesis by blue-green algae, which release oxygen into the air.
1.5 billion–600 million years	Rise of multicellular organisms
545 million years	Cambrian Explosion
500–450 million years	Rise of fish—first vertebrates
430 million years	Algae begin to live on land
420 million years	First appearance of animals on land
350–300 million years	Rise of amphibians
350 million years	Rise of primitive insects
350 million years	Appearance of first plants with roots
300–200 million years	Rise of reptiles
250–205 million years	**Triassic Period**
200–135 million years	**Jurassic Period**
200 million years	Pangaea—the "supercontinent"—begins to break apart
200 million years	Appearance of first mammals
135–65 million years	**Cretaceous Period**
65 million years	K-T Boundary: extinction of dinosaurs and the rise of mammals
65–55 million years	**Paleocene Epoch**
50 million years	Evolution of primitive monkeys

55–38 million years	**Eocene Epoch**
55 million years	Evolution of hares and rabbits
38–26 million years	**Oligocene Epoch**
20 million years	Evolution of parrots and pigeons
26–6 million years	**Miocene Epoch**
20–12 million years	Chimpanzee and hominid lines evolve
6–1.8 million years	**Pliocene Epoch (Quaternary Period)**
4–3 million years	Australopithecus
	A. ramidus
	A. afarensis
	Lucy
	Taunga Child
3–2 million years	*A. africanus*
2.4–1.5 million years	*Homo habilis*
2 million–13,000 years	**Paleolithic (Old Stone) Age: Early hominids develop stone tools**
1.8 million–100,000 years	**Pleistocene Epoch**
1.8 million–100,000 years	*H. erectus*
400,000 years	Archaic *Homo sapiens*
300,000–35,000 years	Neanderthals
120,000 years	Cro-Magnon Man
50,000 years–present	Modern humans (*Homo sapiens*)
13,000–10,000 years	Mesolithic (Middle Stone) Age: Migration of early man accelerates
10,000 years	Neolithic (New Stone) Age: Development of agriculture

Index